Gábor Kenderi

Nonparametric identification of nonlinear dynamic systems

Karlsruher Institut für Technologie
Schriftenreihe des Instituts für Technische Mechanik

Band 32

Eine Übersicht aller bisher in dieser Schriftenreihe erschienenen Bände finden Sie am Ende des Buchs.

Nonparametric identification of nonlinear dynamic systems

by

Gábor Kenderi

Scientific Publishing

Dissertation, Karlsruher Institut für Technologie
KIT-Fakultät für Maschinenbau

Tag der mündlichen Prüfung: 11. November 2016
Referenten: Prof. Dr.-Ing. habil. Alexander Fidlin
Prof. Dr. DSc. Gábor Stépán

Impressum

KIT Scientific Publishing

Karlsruher Institut für Technologie (KIT)
KIT Scientific Publishing
Straße am Forum 2
D-76131 Karlsruhe

www.ksp.kit.edu

Print on Demand 2018 – Gedruckt auf FSC-zertifiziertem Papier

ISSN 1614-3914
ISBN 978-3-7315-0834-2
DOI 10.5445/KSP/1000085419

NONPARAMETRIC IDENTIFICATION OF NONLINEAR DYNAMIC SYSTEMS

Zur Erlangung des akademischen Grades

Doktor der Ingenieurwissenschaften

der Fakultät für Maschinenbau
vom Karlsruher Institut für Technologie (KIT)

genehmigte
Dissertation
von

Gábor Kenderi
aus Budapest

Tag der mündlichen Prüfung: 11.11.2016

Hauptreferent: Prof. Dr.-Ing. habil. Alexander Fidlin

Korreferent: Prof. Dr. DSc. Gábor Stépán

Outline

Identification is an essential part of developing virtual models of dynamic systems. For nonlinear systems the number of describing parameters per degree of freedom and the difficulty of finding the proper describing functions increases with the complexity of the underlying nonlinearities. Therefore, nonparametric identification approaches have some significant advantages over parametric techniques in case of nonlinear systems.

The present thesis proposes a nonparametric identification method for highly nonlinear systems that is able to reconstruct the underlying nonlinearities in form of three-dimensional generalized restoring force surfaces between degrees of freedom, which are arbitrary functions of two state variables, without a priori knowledge of the describing nonlinear functions. There are some known methods for this purpose such as the Restoring Force Surface Method (RFSM) or techniques that utilise the Hilbert-Huang Transform. The current study explores the promising approach of using master-slave synchronisation between the virtual model and the real system, which to the author's knowledge has not yet been investigated in the literature. The synchronization is realized via nonlinear Kalman Filter algorithms, which are optimal observers in a least-squares sense. Using the well-known state augmentation technique, the Kalman Filter can be turned into a dual state and parameter estimator to identify parameters of a priori characterised nonlinearities. The present study proposes an extension of this technique towards nonparametric identification. A general nonlinearity model is introduced by describing the restoring forces via time-variant linear coefficients of the state variables, which are estimated as augmented states. The estimation procedure is followed by an a posteriori probabilistic analysis that reconstructs noisefree restoring force characteristics using the estimated states and their estimated covariance matrices. Observability is provided using only one measured quantity per degree of freedom, which makes this approach less demanding in the number of necessary measurement signals compared with other nonparametric solutions such as the RFSM. Due to the statistical rigour of the procedure, it successfully addresses signals corrupted by significant measurement noise. Thanks to the general Kalman Filter algorithm, the approach can be integrated in a full identification workflow, where the parametric estimation of a priori known functions of the system is carried out as well.

In the present thesis the method is described in detail. This is followed by virtual and real-life identification examples of one and three degree of freedom nonlinear mechanical systems to demonstrate the effectiveness of the proposed technique. Based on these examples several implementation properties of the approach are investigated and advantages as well as challenges in comparison to state of the art methods are discussed.

Abstract

A nonparametric identification method for highly nonlinear systems is presented that is able to reconstruct the underlying nonlinearities in form of three-dimensional generalized restoring force surfaces using vibration measurements without a priori knowledge of the describing nonlinear functions. The approach is based on nonlinear Kalman Filter algorithms using the well-known state augmentation technique that turns the filter into a dual state and parameter estimator, of which an extension towards nonparametric identification is proposed in the present thesis. A general nonlinearity model is introduced by describing the restoring forces via time-variant linear coefficients of the state variables, which are estimated as augmented states. Due to the probabilistic rigour of the procedure, noisefree restoring force characteristics are reconstructed even in the presence of significant measurement noise. Thanks to the Kalman Filter algorithm, observability is provided using only one measured quantity per degree of freedom, and the approach can be integrated in a full identification workflow, where the parametric identification of a priori known functions of the system is carried out as well. The effectiveness of the proposed technique is demonstrated on virtual and real-life identification examples of one and three degree of freedom nonlinear mechanical systems.

Kurzfassung

In der vorliegenden Arbeit wird eine nichtparametrische Identifikationsmethode für stark nichtlineare Systeme entwickelt, welche in der Lage ist, die Nichtlinearitäten basierend auf Schwingungsmessungen in Form von allgemeinen dreidimensionalen Rückstellkraft-Flächen zu rekonstruieren ohne Vorkenntnisse über deren funktionale Form. Die Vorgehensweise basiert auf nichtlinearen Kalman Filter Algorithmen, welche durch Ergänzung des Zustandsvektors in Parameterschätzer verwandelt werden können. In dieser Arbeit wird eine Methode beschrieben, die diese bekannte parametrische Lösung zu einem nichtparametrischen Verfahren weiterentwickelt. Dafür wird ein allgemeines Nichtlinearitätsmodell eingeführt, welches die Rückstellkräfte durch zeitvariable Koeffizienten der Zustandsvariablen beschreibt, die als zusätzliche Zustandsgrößen geschätzt werden. Aufgrund der probabilistischen Formulierung der Methode, können trotz signifikantem Messrauschen störfreie Rückstellkraft-Charakteristiken identifiziert werden. Durch den Kalman Filter Algorithmus ist die Beobachtbarkeit der Nichtlinearitäten bereits durch eine Messgröße pro Systemfreiheitsgrad gegeben. Außerdem ermöglicht diese Beschreibung die Durchführung einer vollständigen Identifikation, wobei die restlichen konstanten Parameter des Systems zusätzlich geschätzt werden. Die Leistungsfähigkeit des entwickelten Verfahrens wird anhand von virtuellen und realen Identifikationsbeispielen nichtlinearer mechanischen Systeme mit ein und drei Freiheitsgraden demonstriert.

Contents

List of figures

Nomenclature

Notations

regular upright	mathematical operations and operators (e.g.: sin(...))
regular italic	scalar values and functions (e.g.: x or $f(x)$)
bold upright	vector or matrix values and functions (e.g.: $\mathbf{x}$ or $\mathbf{f}(\mathbf{x})$)
$\mathbf{P}_\mathbf{X}$	covariance matrix of the random vector variable $\mathbf{X}$ (Exceptions are the process noise vector $\mathbf{q}$ and the measurement noise vector $\mathbf{r}$, whose covariance matrices are denoted by $\mathbf{Q}$ and $\mathbf{R}$ respectively.)
$X \sim (x,P_X)$	random variable X with mean value x and variance P_X
$\mathbf{X} \sim (\mathbf{x},\mathbf{P}_\mathbf{X})$	random vector variable $\mathbf{X}$ with mean vector $\mathbf{x}$ and covariance matrix $\mathbf{P}_\mathbf{X}$
$\mathrm{E}(\mathbf{X})$	expected value of the random vector $\mathbf{X}$
$\hat{\mathbf{x}}$	estimated value of $\mathbf{x}$
$\Phi_\mathbf{h}^k(\mathbf{x},\mathbf{P}_\mathbf{X})$	type k probabilistic transformation of the mean vector $\mathbf{x}$ and covariance matrix $\mathbf{P}_\mathbf{X}$ through the function $\mathbf{h}$
$\mathrm{L}_\mathbf{a}^i(h)$	Lie derivative of order i of the function h with respect to the function $\mathbf{a}$
$\boldsymbol{\mathcal{P}} = \sqrt{\mathbf{P}}$	matrix square-root of the positive definite matrix $\mathbf{P}$ such that $\boldsymbol{\mathcal{P}}\boldsymbol{\mathcal{P}}^\mathrm{T}=\mathbf{P}$

Abbreviations

aRFS	Additive Restoring Force Surface
CMA	Central Moving Average

cRFS	Coupled Restoring Force Surface
DAE	Differential-Algebraic Equation
DDE	Delay Differential Equation
DoF	Degree of Freedom
dRFC	Dissipative Restoring Force Curve
EKF	Extended Kalman Filter
EKS	Extended Kalman Smoother
EMA	Experimental Modal Analysis
eRFC	Elastic Restoring Force Curve
HT	Hilbert Transform
KF	Kalman Filter
KS	Kalman Smoother
ODE	Ordinary Differential Equation
RFC	Restoring Force Curve
RFS	Restoring Force Surface
RFSM	Restoring Force Surface Method
SR-UKF	Square-Root Unscented Kalman Filter
UKF	Unscented Kalman Filter
WTLS	Weighted Total Least Squares

Roman symbols

$\mathbf{a}$, $\mathbf{A}$	process vector function and its Jacobian matrix

$\mathbf{B}$, b	input matrix and its scalar form
c_E , c_D , c	elastic, dissipative and total restoring force function [N]
$\mathbf{c}$	vector of restoring force functions [N]
$\mathbf{d}$, d	instantaneous damping vector and its scalar entries [Nsm^{-1}]
$\mathcal{D}$	linearised damping ratio [-]
e_m , e_p , e_s	measurement, prediction and synchronisation error [-]
e^{WTS}	weighted total squares error [-]
eM	normalised mean error (q.v. Appendix B)
eMS	normalised mean square error (q.v. Appendix B)
$eRMS$	normalised root mean square error (q.v. Appendix B)
eRS	normalised root square error (q.v. Appendix B)
f_E , f_D , f	elastic, dissipative and total restoring force [N]
$\mathbf{h}$, $\mathbf{H}$	measurement vector function and its Jacobian matrix
$\mathbf{I}$	identity matrix
J_ξ	mass moment of inertia with respect to axis ξ [kgm^2]
$\mathbf{K}$	Kalman gain
$\mathbf{K}^{\mathrm{S}}$	smoother gain
$\mathbf{K}^{\mathrm{WTLS}}$	recursive WTLS gain
m	mass [kg]
$\mathbf{M}$	inertia matrix
N	number of DoFs
n	order of a dynamic system, i.e. the number of its state variables

n_u , n_y , n_a	size of the input, measurement and augmented state vectors
n_s	number of samples
$\mathbf{O}$, $\mathbf{o}$	observability matrix and its column vectors
$\mathbf{p}$, p	parameter vector and its scalar entries
p_c	restoring force scaling coefficient [-]
q_s , q_d , q_p	covariance exponents for convergence tuning of **s**, **d** and **p** [-]
$\mathbf{q}$, $\mathbf{Q}$	process noise vector and its covariance matrix
$\mathbf{r}$, $\mathbf{R}$	measurement noise vector and its covariance matrix
$\mathbf{s}$, s	instantaneous stiffness vector and its scalar entries [Nm^{-1}]
t	time [s]
T_{NC}	normed CPU time [-]
$\mathbf{u}$, u	input vector and its scalar entries
$\mathbf{v}$, v	rate of deformation vector and its scalar entries [ms^{-1}]
$\mathbf{w}$	N-element set of selected state space variables and derivatives
$\mathbf{x}$, $\mathbf{x}_a$	state vector and its augmented version
$\mathbf{y}$, y	measurement vector and its scalar entries
$\mathbf{z}$, z	deformation vector and its scalar entries [m]
$\mathbf{0}$	zero vector or zero matrix

Greek symbols

α	undamped natural angular frequency [s^{-1}]
β_j	arbitrary function of **x** defining z_j

γ_j	arbitrary function of $\mathbf{x}$ defining v_j
δ	mass-normalised damping coefficient [s^{-1}]
$\boldsymbol{\kappa}$	state constraint vector function
$\boldsymbol{\lambda}^{\mathrm{RFC}}$	vector function of $\mathbf{X}_C$ that defines RFC samples
$\boldsymbol{\lambda}^{\mathrm{RFS}}$	vector function of $\mathbf{X}_{\mathrm{RFC}}$ that defines RFS samples
μ_{ij}	entries of the inverse inertia matrix $\mathbf{M}^{-1}$
Λ	instantaneous amplitude [m]
ω	instantaneous angular frequency [s^{-1}]
Ψ	instantaneous phase [rad]
φ_ξ	rotation coordinate around axis ξ [rad]
φ_η	rotation coordinate around axis η [rad]
$\boldsymbol{\rho}_E$	sample of the uncertainty curve of a fitted eRFC
$\boldsymbol{\rho}_D$	sample of the uncertainty curve of a fitted dRFC
$\boldsymbol{\sigma}$	RFS sample vector
$\boldsymbol{\sigma}_E$	eRFC sample vector
$\boldsymbol{\sigma}_D$	dRFC sample vector
τ	time constant [s]
τ_{D}	time delay [s]
θ	arbitrary phase shift [rad]
ξ	first axis of a Cartesian coordinate system in 3D
η	second axis of a Cartesian coordinate system in 3D
ζ	third axis of a Cartesian coordinate system in 3D

Foreword

This thesis was accomplished during my work at LuK GmbH & Co KG in Bühl under the supervision of the Institute of Engineering Mechanics (Institut für Technische Mechanik, ITM), Department of Mechanical Engineering at the Karlsruhe Institute of Technology (KIT). I would like to thank all the colleagues at LuK and at KIT for their contributions to my work, and I would like to express, how grateful I am for their support.

First of all, I would like to thank my supervisor, Prof. Dr.-Ing. habil. Alexander Fidlin, for his continuous support with my work. He came up with the idea of exploring the possibilities in synchronisation based identification, which led to five years full of exciting scientific challenges for me. In my eyes he is not "only" an excellent scientist and researcher but also someone who pays great attention to motivate and educate future generations.

I would like to express my thanks to LuK GmbH & Co KG for making this study possible. Due to its great engineering know-how and the innovative atmosphere at work, it is a perfect place for a doctoral research.

Special thanks to Prof. Dr. DSc. Gábor Stépán for all the effort he invested as the second reviewer. I must take this opportunity to also thank him for his extremely inspiring lectures during my studies in Budapest. It was his influence that turned my attention to applied mechanics.

I would like to thank Prof. Dr.-Ing. Carsten Proppe and Prof. Dr.-Ing. Wolfgang Seemann for welcoming me as a Ph.D. student at the Institute of Engineering Mechanics.

I would like to express many thanks to all my colleagues at LuK and the employees at KIT for all the time we spent together. I would like to thank Dr. Hartmut Faust for his useful remarks on my thesis. Special thanks goes to Dirk Klünder for spending plenty of his time sharing his expert knowledge in vibration testing with me. Thanks to Arsenty Tikhomolov, Laurent Ineichen and Jens Deppler for helping me in all possible challenges of accomplishing a doctoral thesis. I would also like to thank Georg Jehle and Marius Köster for the interesting discussions on system identification, and Hartmut Hetzler, from whom I learned a lot during our experiments in ITM's laboratory.

Thanks to all my friends for accepting my Ph.D. as an excuse so many times, and for making me laugh, when I was feeling down. I would like to thank Michael for spontaneously offering to check my thesis for mistakes. Thank You, Zsolt and Jens, for always being there for me.

I would like to thank my father for constantly showing me, what it means to be a good engineer, and my mother for always telling me, what I need to hear instead of what I want to hear. I would like to thank my grandparents for all those wonderful childhood memories, and my sister for knowing me better than I know myself and always keeping an eye on me. I am very grateful to be a part of this family.

At last but not at least: Thank You, my Katharina, for giving my life a purpose.

Gábor Kenderi

Karlsruhe, August 2016

1 Introduction and literature overview

1.1 Motivation

Virtual models and numerical simulation have become crucial parts of the development of nearly all human-made structures. Their importance is ever growing as the computational power of modern computers is increasing exponentially and the related hardware costs are decreasing rapidly. Software development of the past few decades led to a wide variety of simulation environments that enable the engineers of today to model complex dynamic systems (Dresig & Fidlin, 2014). This development makes virtual models highly attractive to replace costly physical models and experiments. Such models are however not ready to be employed for the prediction of the real system's behaviour until their properties (i.e. parameters and characteristics) have been tuned to appropriate values. Therefore, system identification is an essential part of developing virtual models (Kerschen, Worden, Vakakis, & Golinval, 2006), (Kerschen, Worden, Vakakis, & Golinval, 2007).

For linear systems well-defined frameworks provide straightforward methodologies to solve the identification problem (Ljung, 2013), (Okuma & Oho, 1997). Linear consideration of dynamic systems is very popular. It is and most certainly will remain an important domain of virtual model development. However, "most of everyday life is nonlinear" (Strogatz, 1994), since "the world around us and we ourselves are inherently nonlinear" (Fidlin, 2006). Simply-sounding properties of dynamic systems such as the presence of friction or end-stop (Fidlin, 2006) or time-delay (Stépán, 1989) can lead to extremely complex and exotic behaviour. Unfortunately there are no general identification solutions for nonlinear systems. Numerous approaches are available, each of them having their advantages and drawbacks depending on the specific system of interest.

There are some general requirements that should be fulfiled by an identification procedure:

- The necessary measurement setup should be as simple as possible. (Decreasing experiment complexity to develop less costly methodologies for determining system properties is an important trend in test rig development (Klöpper, Okuma, & Krüger, 2013).)
- The procedure should be robust against measurement noise. If possible the engineer should receive a quantification of the uncertainty of the identified properties caused by the noisiness of the measurement signal. (Although the accuracy of available experimental equipment is increasing rapidly, measurement noise is still unavoidable and in most situations can not be neglected.)
- The procedure should be computationally as fast and simple as possible. (Although the capacity of today's computers is increasing fast, the demand on fast and flexible product development is increasing as well.)

In comparison to linear systems there are some additional difficulties arising from nonlinearities:

- The number of system describing parameters per DoF increases leading to higher computational demand and sometimes convergence problems during the identification.
- The recognition of the type of nonlinearities in order to define the system parameters prior to the identification procedure is in some cases very difficult.

- Nonlinear phenomena (such as bifurcations, limit cycles and chaotic behaviour) are highly sensitive to model parameters, which leads to an increased necessity of identification accuracy.

Under these conditions nonparametric identification approaches (q.v. Section 1.4.2) have some significant advantages over parametric techniques (q.v. Section 1.4.1), since for a wide class of nonlinearities (q.v. Section 1.2) they do not require the a priori definition of the nonlinear functions. Hence, there is a great need of nonparametric identification methods that fulfil the aforementioned requirements.

1.2 Nonlinearities in dynamic systems

A general nonlinear system, whose dynamics are governed by ordinary differential equations (ODEs), is given by two vector functions (**a** and **h**) of the state vector **x** and the input vector **u** that define the system consisting of the process equation (1.1) and the measurement equation (1.2) of the form

$$\dot{\mathbf{x}} = \mathbf{a}(\mathbf{x}, \mathbf{u}, t), \tag{1.1}$$

$$\mathbf{y} = \mathbf{h}(\mathbf{x}, \mathbf{u}), \tag{1.2}$$

where **y** denotes the vector of measured quantities with a size of n_y, which defines the number of measurement signals. Let the number of states (the size of **x**) and the number of inputs (the size of **u**) be denoted by n and n_u respectively. Processes given by differential-algebraic equations (DAEs) and delay differential equations (DDEs) (Stépán, 1989) are not considered explicitly at this point. However, in Section 1.6.3 and 2.2.3 it is shown that the presented procedure can accommodate DAEs as a general form of nonlinearity and at the end of Section 3.4.1 the possibilities of extending the approach towards DDEs are briefly discussed.

In most cases it is possible to separate a set of time-invariant functions of the state variables, denoted by the vector of functions **c**, inside the system equations, which are then given as

$$\dot{\mathbf{x}} = \mathbf{a}(\mathbf{x}, \mathbf{u}, \mathbf{c}(\mathbf{x}), t), \tag{1.3}$$

$$\mathbf{y} = \mathbf{h}(\mathbf{x}, \mathbf{u}, \mathbf{c}(\mathbf{x})). \tag{1.4}$$

In the special case, when each of these functions, denoted by c_j, are two-dimensional (i.e. their values depend on two independent variables, denoted by z_j and v_j), the vector **c** has the form

$$\mathbf{c} = \left[c_1(z_1, v_1), \dots, c_j\left(z_j, v_j\right)\right]^{\mathrm{T}}, \tag{1.5}$$

where each independent variable is an arbitrary function of the state variables:

$$z_j = \beta_j(\mathbf{x}), \tag{1.6}$$

$$v_j = \gamma_j(\mathbf{x}). \tag{1.7}$$

Such functions can not describe explicitly time-dependent sources of nonlinearities, e.g. parametric excitation (Insperger & Stépán, 2002), however, a huge domain of nonlinearities are covered by this formulation. In the physical domain of mechanical systems such functions mostly represent general restoring forces (force or torque) between DoFs of the system. These functions are called force-state maps (Crawley & Aubert, 1986) and their representing three-dimensional surfaces are called restoring force surfaces (Kerschen, Worden, Vakakis, & Golinval, 2006), (Link, Boeswald, Laborde, Weiland, & Calvi, 2011), denoted by RFS. If e.g. z_j and v_j are two relative displacements inside the system, then RFS_j (defined by c_j) represents a nonlinear coupled stiffness characteristic. Such characteristics are used e.g. to model nonlinear clutch components (Tikhomolov, 2015). However, the most typical RFSs represent nonlinear spring-dampers, where z_j is the deformation and v_j is the rate of deformation of the spring-damper. A well known example is a Liénard system (Strogatz, 1994), where the RFS is of the form

$$c(z, v) = d(z)v + s(z)z. \tag{1.8}$$

In this particular RFS a coupled (z- and v-dependent) term is to be found. Such RFSs are called a "coupled RFS", denoted by cRFS. On the other hand if an RFS can be given as the sum of a z-dependant "elastic" part and a v-dependant "dissipative" part, it is called an additive RFS, denoted by aRFS, which has the form

$$c(z, v) = c_E(z) + c_D(v). \tag{1.9}$$

Let c_E and c_D be called elastic and dissipative restoring force curves with the abbreviations eRFC and dRFC respectively. If possible, one tries to model systems with additive RFSs, since their analytical investigation is less complex. Nonlinearities such as progressive and degressive stiffness, backlash, end-stop and friction can be described by such models. Many experimental, analytical and numerical investigations of such systems can be found in the literature, e.g. (Ineichen, 2013), (Tikhomolov, 2015), (Ing, Pavlovskaia, & Wiercigroch, 2011), (Viguie & Kerschen, 2010). For an extensive overview of such nonlinearities, incl. their advantages and undesirable effects in engineering structures, the reader is referred to (Fidlin, 2006), (Dresig & Fidlin, 2014) and (Ibrahim, 2008).

The importance of RFSs lies in the fact that they can be identified in a nonparametric way, to which the present study aims to offer a synchronisation-based solution. Although the term RFS originates from the physical domain of mechanics and the current thesis concentrates on the identification of mechanical systems, no restrictions will be made that would not allow one to transfer the investigated methods to other physical domains. (E.g. in Section 3.2 the identification of a Van der Pol system is taken as an example that originates from electronics.)

1.3 Problem formulation

For the derivation of the synchronisation-based nonparametric identification procedure a particular class of mechanical structures is considered that represents a special subclass of the systems defined by (1.3) and (1.4). Except for the excitation (input) **u** the system has no explicit time-dependence. The considered structures have N DoFs connected via N RFSs in a way that the deformation and the rate of deformation coordinates of these RFSs (i.e. z_j and v_j) can be defined as state variables, which leads to a state vector of length $2N$ in the form

$$\mathbf{x} = \begin{bmatrix} \mathbf{z}_{N\times 1} \\ \mathbf{v}_{N\times 1} \end{bmatrix}, \tag{1.10}$$

where **z** is the vector of the deformation coordinates and **v** the vector of the rate of deformation coordinates of the N RFSs respectively. In the current study many load cases will be presented, where the excitation **u** is of kinematic type. In such situations the above choice of state space coordinates is advantageous, since no integrals or derivatives of the time series of **u** will appear in the system equations. The above assumptions lead to the following particular form of the process equation:

$$\begin{bmatrix} \dot{\mathbf{z}}_{N\times 1} \\ \dot{\mathbf{v}}_{N\times 1} \end{bmatrix} = \begin{bmatrix} \mathbf{v}_{N\times 1} \\ -\mathbf{M}^{-1}_{N\times N} \begin{bmatrix} c_1(z_1, v_1) \\ \vdots \\ c_N(z_N, v_N) \end{bmatrix}_{N\times 1} + \mathbf{B}_{N\times n_\mathrm{u}} \mathbf{u}_{n_\mathrm{u}\times 1} \end{bmatrix}, \tag{1.11}$$

where **M** denotes the inertia matrix. The input matrix **B** defines how the excitation **u** is acting on the mechanical system. If the system is excited by forces, then the input matrix consists of specific elements of the inverse inertia matrix. In case of a kinematic excitation in the form of acceleration the elements of **B** directly assign the components of **u** to the corresponding rows of equation (1.11), i.e. $B_{ij} \in \{-1,0,1\}$.

The primary aim of the present study is to investigate a nonparametric identification method based on (1.4) and (1.11) that reconstructs the N RFSs (i.e. the functions c_j) based on the time history of **u** and noisy measurements of **y**. Throughout the derivation of the nonparametric approach in Chapter 2 the matrix **M** (and therefore **B** also) is assumed to be known. This assumption is relaxed later in Chapter 4, where the additional parametric identification of **M** is investigated.

1.4 Identification methods of nonlinear systems

In (Kerschen, Worden, Vakakis, & Golinval, 2007) the identification of nonlinearities is defined to consist of three phases: detection, characterisation and parameter identification. Possible solutions to the detection problem are outside the scope of the current thesis. Answering the question, whether the system of interest exhibits a nonlinear behaviour that can not be neglected (linearised), i.e. the detection, is considered to be already done. We assume the case, where nonlinear identification is necessary. Therefore, only characterisation and parameter identification are discussed in the present study. Still it can be mentioned that also in situations, where detection results are uncertain or simply not available, it is safer to start with nonlinear methods, since linear behaviour is a special case of nonlinearity. Although the current thesis is dedicated to explore a nonparametric technique that allows the resignation of the parameter identification phase, it is important to point out that a method containing this phase can by definition still be a nonparametric approach. This will become clear later on in this section.

The characterisation of mechanical systems described by (1.4) and (1.11) requires the approximation of the RFSs by a particular function of the form

$$c_j(z_j, v_j) \approx \hat{c}_j(z_j, v_j, \mathbf{p}_j), \tag{1.12}$$

where $\mathbf{p}_j$ denotes a vector of parameters that is to be estimated during the parameter identification phase. The purpose of system identification is to extract the highest possible amount of global information from the measurement signals and to assign the gained information to the local system properties. This always implies the assumption of an a priori model of the system. The procedure is then carried out by fitting this model to the measurements in the time- or frequency-domain using least-squares, maximum likelihood, nonlinear optimization approaches or explicit formulas. Whether the a priori model requires the characterisation of the nonlinearities, i.e. the definition of the approximating functions (1.12), or not is an essential property of nonlinear identification techniques that leads to their classification into two groups: "parametric" or "nonparametric". In the following a brief overview of such techniques for nonlinear systems is given in order to place the investigated approach in the multitude of existing methods.

1.4.1 Parametric approaches

In these techniques, the characterisation of the RFSs is included in the a priori model assumptions. In this case, the fusion of the a priori model with the measurement data, i.e. the fitting procedure, directly estimates the best fitting values of $\{\mathbf{p}_1, \mathbf{p}_2, \dots, \mathbf{p}_N\}$.

Optimisation based approaches

Possibly the most straightforward way of nonlinear identification is to choose an appropriate function of the parameter vectors, denoted by e_O, and use a suitable optimisation algorithm to find its global minimum. This function is called the "objective function" that should represent the error between the real system and the model. Typically the mean square (MS) error is chosen for this purpose, which leads to

$$e_O(\mathbf{p}_1, \mathbf{p}_2, \ldots, \mathbf{p}_N) = \sum_{i=1}^{n_S} (\mathbf{y}_i - \hat{\mathbf{y}}_i)^T (\mathbf{y}_i - \hat{\mathbf{y}}_i)\,, \tag{1.13}$$

where $\hat{\mathbf{y}}$ denotes the estimate of the measurement generated by the model and n_S is the number of measurement samples. This approach is presented in (Kunath, Niemeier, Schlegel, & Will, 2014) in its classical form, the so called Single Shooting. In this approach a complete simulation of the model is done at each iteration step using the whole time series of $\mathbf{u}$ and one set of parameters. e_O is only analysed between the simulation runs. This formulation often results in objective functions with several local minima, which implies the use of global optimisation techniques such as genetic algorithms (Charbonneau, 2002). Such solutions have the drawback that they can easily end up in an enormous number of iterations, which often leads to unacceptable identification time. Another difficulty of single shooting is that the initial error is cumulated throughout the whole integration, which can even result in failed simulation runs for "bad" parameter sets (Michalik, Hannemann, & Marquardt, 2009). These difficulties motivated several techniques that aim to form e_O into a function with one (global) minimum. In this case much faster local optimisation algorithms can be applied such as the derivative based Newton's method (Nocedal & Wright, 1999) or the bit slower but derivative free Simplex algorithm (Charbonneau, 2002), (Lagarias, Reeds, Wright, & Wright, 1998).

One approach dedicated to avoid several local minima is called Multiple Shooting (Voss, Timmer, & Kurths, 2004). Here the measurement time is split into multiple simulation runs. The initial states of the time segments are introduced as additional parameters, which increases the number of parameters to be optimised. However, due to small integration periods the estimated measurement is always kept near to the measured signals, which results in a beneficial objective function. The Incremental Single Shooting (ISS), proposed in (Michalik, Hannemann, & Marquardt, 2009), aims to combine the advantages of both the Single and the Multiple Shooting technique.

Another solution of this kind is the Homotopy method that modifies the system model into an observer of the form

$$\dot{\hat{\mathbf{x}}} = \mathbf{a}(\hat{\mathbf{x}}, \mathbf{u}, \hat{\mathbf{p}}_1, \hat{\mathbf{p}}_2, \dots, \hat{\mathbf{p}}_N) + \mathbf{K}(\mathbf{y} - \hat{\mathbf{y}}), \tag{1.14}$$

Where the "hat" symbol denotes the estimates of the specific quantities and **K** denotes the so called "synchronisation gain" that recursively forces the model to stay near to the measured behaviour. Successful implementation of this technique for multiple DoF vehicle models is presented in (Vyasarayani, Uchida, Carvalho, & McPhee, 2012) and the application to identify nonlinear oscillators and a clutch actuation system with nonlinear characteristics is investigated in (Gunnarsson, 2014). In these works a multi-step version of the approach is used, where multiple optimisation steps are carried out by reducing the synchronisation gain from step to step until the final result is reached. In (Sun & Yang, 2010) and (Carlsson & Nordheim, 2011) a promising single-step version of this idea is implemented. For the calculation of **K** the recursive optimal approach of Unscented Kalman Filtering (UKF) is used and the classical e_O from equation (1.13) is replaced by the maximum likelihood function. These modifications aim to result in the optimal solution in the first step of the homotopy algorithm making the further iteration steps with decreasing **K** unnecessary.

Frequency domain parametric approaches

Frequency domain techniques, such as Experimental Modal Analysis (EMA), provide perhaps the most convenient way to identify linear systems. One of their important advantages is their ability to reconstruct the complete underlying linear system under specific measurement conditions without the a priori knowledge of the inertia matrix **M**. See e.g. in (Kletschkowski, 2013), where acoustic measurements are used for this purpose or in (Richardson, 1977), where the general theory of EMA based linear system identification is summarised. The extension of these techniques towards nonlinear systems motivated several investigations with fruitful results. Different approaches are discussed in (Worden & Tomlinson, 2001) that aim to adapt the theory of EMA to nonlinearities. In (Peeters, 2010) an extensive study on the usefulness of Nonlinear Normal Modes (NNM) in the investigation of nonlinear systems is presented. The same theory is applied in (Laxalde & Thouverez, 2009) for the modal investigation of turbo machinery blades. In (Platten, Wright, Cooper, & Sarmast, 2002) the underlying multiple DoF linear system is identified at low excitation level. Then at high excitation level the nonlinear modes are detected and subsequently excited by appropriate force vectors in order to separately identify the nonlinearities of each mode in frequency domain. Mode couplings are also identified using the Nonlinear Resonant Decay Method (NL-RDM). Similarly in (Zanotti Fragonara, et al., 2012) well separated nonlinear modes of a bridge are identified considering one modal DoF at a time. The Frequency-domain Nonlinear Subspace Identification method (FNSI) is successfully applied to a high-dimensional nonlinear real-life structure in (Noël & Kerschen, 2013).

Another powerful tool for nonlinear parameter identification in frequency domain is the Hilbert-Huang Transform (HHT), which approaches the problem from a different point of view than the previous techniques. It can be classified as time-frequency method (Kerschen, Worden, Vakakis, & Golinval, 2006), since it reconstructs time-varying frequency domain properties of measured signals. It is based on the Hilbert Transform (HT) that enables the determination of the instantaneous amplitude, phase and frequency of monocomponent oscillatory signals as time-series. The HHT extends this algorithm to multicomponent signals using the empirical mode decomposition (EMD) to extract the periodic components (Barnhart, 2011). Based on the results of the HHT different approaches have been proposed that allow the parametric identification of multiple DoF nonlinear systems. One of these algorithms, the so called slow-flow model identification (SFMI), is proposed in (Kerschen, Vakakis, Lee, McFarland, & Bergman, 2008).

Time-series analysis

In time-series analysis the system model is formulated in difference equations instead of ODEs. The goal of the procedure is to identify the coefficients of these equations by fitting in time domain. In (Gelb, Kasper, Nash, Price, & Sutherland, 2001) the basic models for linear systems are derived. A very effective extension of the approach to nonlinear systems, called NARMAX modelling, is presented in (Kerschen, Worden, Vakakis, & Golinval, 2006).

Observer-based methods

Observer-based methods include stochastic approximation discussed in (Gelb, Kasper, Nash, Price, & Sutherland, 2001) and optimal estimation algorithms such as the Kalman Filter (KF). These methods rely on the modification of the process equation into (1.14). The basic idea that enables this state observer to be applied for parameter identification is the so called state augmentation technique, where unknown parameters are introduced as additional state space coordinates, which leads to the so called dual state and parameter estimation. This approach requires nonlinear extensions of the Kalman Filter such as the Extended KF (EKF), the Unscented KF (UKF) or the Particle Filter (PF). See e.g. in (Simon, 2006). Since this approach plays an essential role in the current study, a more detailed description of these algorithms is given in Section 1.6. In this section the attention is paid to successful applications in the literature. The dual state and parameter estimation of mechanical structures is carried out in (Wu & Smyth, 2007), (Chatzi & Smyth, 2009) and (Li, Suzuki, & Noori, 2004) to estimate coefficients of strongly nonlinear hysteretic characteristics, and in (Kolansky & Sandu, 2012) to estimate inertia properties of ground vehicles. In

(Bessa, Hackbarth, Kreuzer, & Radisch, 2014) the identification of an electro-hydraulic servo system is carried out and the implementation of the augmented UKF is presented in (Sitz, Schwarz, Kurths, & Voss, 2002) on the example of various academic nonlinear systems exhibiting limit cycles and chaotic behaviour. A robust modification of the KF, the H∞ filter, is proposed in (Kiriakidis & O'Brien Jr., 2004) to handle uncertainties in the a priori model. A globally iterated KF can be used to overcome the problem of insufficient convergence of parameters e.g. due to short-time measurement signals. This idea is investigated in (Hoshiya & Saito, 1984) and in (Voss, Timmer, & Kurths, 2004). In the latter the performance of the technique is shown to be similar compared to Multiple Shooting.

DoF identification

A special parametric identification method that should be mentioned is the so called Phase Space Reconstruction. Its aim is outside the scope of the problem formulated in Chapter 1.3, since it deals with the identification of the fundamental system property of the number of DoFs, i.e. N, based on measurements of nonlinear attractors of the system. An explanation of the method including some academic examples can be found in (Prahs, 2011).

Estimation of the time history of varying parameters

Some of the aforementioned methods, namely the ones that give parameter estimates for each measurement time step, enable the estimation of the time history of varying parameters, i.e. the identification of time-variant systems. These methods represent a bridge to nonparametric identification, which is discussed in the next section.

The reconstruction of the excitation time-series acting on the system, denoted by $\mathbf{u}$ in equation (1.1), based on a priori system models and measured system responses is an important field of system monitoring, see e.g. (Sturm,

Moorhouse, Kropp, & Alber, 2013). This topic is outside the scope of the current study, however, many of the techniques applied to solve this problem can straightforwardly be applied to identify time-series of varying parameters as well. The previously mentioned KF is often used for this purpose in its augmented form with **u** introduced as state variable, since it gives an estimate of the model states recursively at every time step. This technique is applied in (Zeile & Maione, 2015), (Lourens, Reynders, De Roeck, Degrande, & Lombaert, 2012) and (Lourens, et al., 2012) for monitoring of system loads based on response measurements. In such applications of the KF a new challenge arises regarding the convergence of the augmented states. In this case it is not enough to get a converged parameter value at the end of the measurement sequence. The convergence has to be fast enough to track the time-variance, which on the other hand leads to increased noisiness of the results, since the observer somewhat starts to follow the measurement noise as well. This trade-off between bias (phase shift) and variance (noisiness) is discussed in (Hansen, 1992) by means of the L-curve in general mathematical form. The L-curve point of view has found its application in the above mentioned papers as well. If a priori knowledge about the expected form of the time-variance is available, this can be integrated in the estimation procedure in order to reduce the phase shift without increasing the noisiness of the estimation results. This idea is used in (Jakubek & Fleck, 2009) for the estimation of combustion engine inner torque using an augmented Kalman Filter. Here the estimation of rapidly varying time signals is replaced by the estimation of constant or slowly varying coefficients of specially designed basis functions. If the mentioned a priori knowledge is not available one can introduce additional state variables, which represent time derivatives of the augmented state that we are actually interested in. This can also drastically reduce phase shift, however, on the cost of increasing observer complexity. These methods originate from the theory of target tracking and are discussed in (Simon, 2006) and (Wu, Lin, Han, & Li, 2010). The resulting dynamic models of the augmented states are often referred to as Random Walk (RW) models (Young, 2000).

The capability of the KF to identify time-varying systems enables its on-line (real time) implementation. Such an application is presented in (Garcia & Antsaklis, 2009), where an augmented EKF is integrated in the algorithm of a model based control circuit in order to update the virtual model parameters in real time.

The aforementioned HHT approach also enables the estimation of time-series of system parameters. Such a procedure is investigated in (Hu & Proppe, 2011) and (Hu & Proppe, 2012), where nonlinear time-varying 1DoF and 2DoF systems are identified based on noisy measurements.

General parametric functions

An extension of parametric techniques for cases, where a priori characterisation is not possible due to a lack of information, is the introduction of general parametric approximations of the RFSs instead of using physical parameters. This can, for instance, require the application of power series polynomials, splines, Chebyshev series or Volterra series. An implementation of the latter in frequency domain is presented e.g. in (Németh, Kollár, & Schoukens, 2001). Confusingly this extension is often called "nonparametric". This is the case in (Noël & Kerschen, 2013), where cubic splines are used, and also in (Masri, Chassiakos, & Caughey, 1992), where weighting parameters of a neural network are identified. Technically, these solutions are still parametric, since the applied general functions are indeed a form of a priori characterisation. Here we pay with increased model complexity for not having to deal with the characterisation for specific applications. Because of this, the actual nonparametric solutions are called "truly nonparametric" in (Kerschen, Worden, Vakakis, & Golinval, 2006) for the sake of distinguishability.

Summary of parametric methods

Although parametric approaches are powerful solutions that can effectively address high-dimensional problems, they still have one drawback: an a priori knowledge of the nonlinearities is needed to achieve a quality result without computational problems. Using general mathematical models that can describe arbitrary nonlinearities requires a large number of parameters that can lead to difficulties during the identification process. Further specific problems can occur in case of strong nonlinearities. If the nonlinear function is e.g. piecewise continuous, identifyability problems can arise. Such a difficulty is revealed in (Bessa, Hackbarth, Kreuzer, & Radisch, 2014).

1.4.2 Nonparametric approaches

In comparison to the previously discussed techniques, in nonparametric approaches the a priori model assumptions do not include the characterisation of the RFSs, i.e. (1.12). While the measurement signals are processed, a point cloud of samples is generated that represents the RFSs of the system. Depending on the specific technique these samples can represent different measurement load cases or different time instances of the same measurement time-series. For cRFSs the ith sample of the jth RFS is given by a coordinate triplet $\{z_j,v_j,f_j\}_i$ with f_j denoting the jth restoring force. In case of an aRFS two separate coordinate pairs, $\{z_j,f_{E,j}\}_i$ and $\{v_j,f_{D,j}\}_i$, are generated representing the jth eRFC and dRFC respectively, where $f_{E,j}$ denotes the jth elastic and $f_{D,j}$ the jth dissipative restoring force. From (1.9) it follows that

$$f_{j,i} = f_{E,j,i} + f_{D,j,i}. \tag{1.15}$$

The local characterisation and parameter estimation can be performed a posteriori using the generated sampling points, which is the main practical advantage of nonparametric methods. During the fusion of the a priori model with the measurements only a few fitting parameters are introduced, also known as "instrumental variables", which represent the coordinates of the RFS samples or some other parameters directly related to them (e.g. instantaneous stiffness and damping coefficients).

It is important to distinguish this definition from another wide-spread interpretation of "nonparametric" system representation, where the identification consists of capturing frequency response, step response or impulse response functions by a set of response coefficients, instead of describing the process in a physical form (Wang, Gawthrop, & Young, 2005).

Frequency domain nonparametric methods

A group of nonparametric identification methods is based on frequency domain a priori models with time-varying or load-case-varying modal parameters. For a 1DoF autonomous system this leads to a process equation of the form

$$\ddot{z} + \delta(i)\dot{z} + \alpha^2(i)z = 0, \tag{1.16}$$

where $\delta(i)$ represents a mass-normalised varying damping coefficient and $\alpha(i)$ denotes the varying undamped natural angular frequency of the oscillator. The measured time-domain quantities are also transformed into frequency domain quantities, such as frequency and amplitude, and so the fitting is carried out in the frequency domain. In the 1DoF case the modal parameter samples can be transformed a posteriori into RFS samples by explicit formulas. E.g. using the mass of the oscillator a sample of α can be transformed into an effective stiffness sample, which can be further transformed into f_E using the current vibration amplitude. From this it follows that such approaches imply the a priori knowledge of the oscillating mass. The main restriction of these techniques is that for time domain models they can only be applied in case of 1DoF systems. It is possible to identify modal RFSs of multiple DoF systems if they only exhibit well separated resonance ranges without mode coupling effects. However, these modal RFSs can not be transformed to physical coordinates, and therefore only a frequency domain nonlinear model can be established. On the other hand working in frequency domain leads to two major advantages. The first one is that one measured quantity (e.g. displacement) is enough for the identification of an RFS, since the others (e.g. velocity and acceleration) can be calculated analytically using simple laws of harmonic oscillations. The second one is the good robustness against measurement noise.

In (Link, Boeswald, Laborde, Weiland, & Calvi, 2011) an approach of this kind is investigated, where the different RFS samples belong to separate experimental modal analysis (EMA) load cases with different vibration amplitudes. The method is based on the quasi-linearisation of nonlinear systems in a particular operating point. Since the EMA is an experimental methodology for linear systems, its application to a strongly nonlinear frequency response is not trivial. As a solution to this problem the so called Controlled Response technique is presented in the above paper. In this technique the excitation signal is specially controlled during the experiment, which results in linear resonance characteristics. Due to this the nonlinearity is only observable as the shifting of the quasi-linear eigenfrequency depending on the different vibration amplitudes between different test runs. The related theory of the so called Describing Functions, which approximately describe the transfer characteristics of nonlinearities, can be found in (Gelb & Van der Velde, 1968).

It has been mentioned in Section 1.4.1 that the Hilbert Transform (HT) enables the identification of the time series of time-varying parameters. This property has the fruitful outcome that the HT approach can be formulated in a nonparametric form as well if the time-varying values are linked together to form RFS samples. This is proposed in (Feldman, 1994) for forced vibration experiments and in (Feldman, 1997) for measurements of free oscillations in a generalised form for strong nonlinearities with fastly varying frequency domain character. In (Kerschen, Worden, Vakakis, & Golinval, 2006) the basic equations of this algorithm are summarised, which are briefly reviewed in the following. Consider the 1DoF system from (1.16). Suppose that we directly measure the displacement of the system during free oscillation, and that this time series can be described in form of a monocomponent harmonic signal with time-varying parameters. The oscillation is then defined by

$$z(t) = \Lambda(t)\cos\big(\Psi(t)\big) \quad \text{and} \quad \tilde{z}(t) = \Lambda(t)\sin\big(\Psi(t)\big), \tag{1.17}$$

where $\tilde{z}(t)$ is the Hilbert Transform of $z(t)$. $\Lambda(t)$ and $\Psi(t)$ denote the instantaneous amplitude and instantaneous phase respectively that can both be analytically calculated using the equations

$$\Lambda(t) = \sqrt{z^2 + \tilde{z}^2} \quad \text{and} \quad \Psi(t) = \tan^{-1}\left(\frac{\tilde{z}}{z}\right). \tag{1.18}$$

The instantaneous angular frequency, denoted by $\omega(t)$, is defined analytically as well by the expression

$$\omega(t) = \dot{\Psi}(t) = \frac{z\dot{\tilde{z}} - \dot{z}\tilde{z}}{z^2 + \tilde{z}^2}. \tag{1.19}$$

Using these quantities the time-varying coefficients from (1.16) can be calculated by the equations

$$\delta(t) = -2\frac{\dot{\Lambda}}{\Lambda} - \frac{\dot{\omega}}{\omega} \quad \text{and} \quad \alpha^2(t) = \omega^2 - \frac{\ddot{\Lambda}}{\Lambda} - \delta\frac{\dot{\Lambda}}{\Lambda}. \tag{1.20}$$

This finally leads to the instantaneous samples of the elastic and dissipative restoring force in the form

$$f_E(t) = \Lambda m\alpha^2 \quad \text{and} \quad f_D(t) = \omega\Lambda m\delta, \tag{1.21}$$

where m denotes the mass of the oscillator that, as previously mentioned, has to be available a priori. The forced vibration version of the HT approach (Feldman, 1994) is in contrast to the above discussed free vibration method restricted to symmetric nonlinearities. For general nonlinearities the HT approach is only applicable if free oscillation measurements are available.

Restoring Force Surface Method

The idea of the Restoring Force Surface Method (RFSM) is very simple and intuitive. Based on equation (1.11) the restoring forces, denoted by $f_j(t)$ for the jth DoF, can explicitly be calculated at every measurement time step using the formula

$$\left[f_j(t)\right]_{N\times 1} = -\mathbf{M}\dot{\mathbf{v}} + \mathbf{M}\mathbf{B}\mathbf{u}\,. \tag{1.22}$$

Using the time signal of the state vector **x**, RFS sampling points can be directly generated for every time step. Due to its simplicity and the fact that it introduces no restrictions regarding the RFSs that are identifiable, this method has found numerous implementations in the literature. In (Crawley & Aubert, 1986) and (Peifer, Timmer, & Voss, 2003) the RFSM based identification of 1DoF structures is presented. The latter paper also addresses the optimal choice of smoothing for the a posteriori parametric fitting of the RFS. The errors-in-variables (EIV) problem, that arises from the noisiness of the measured **x**, is mentioned as well, however, without taking it into account in the analysis. The RFSM is used for the identification of a nonlinear wire rope in (Kerschen, Lenaerts, & Golinval, 2001). In (Park & Kim, 1994) the performance of the RFSM is compared to a frequency domain method based on substructuring for two mechanical structures including Coulomb friction. The method is applied for a multiple DoF spacecraft in (Noël, Kerschen, & Newerla, 2012), where the underlying linear system is identified prior to the RFSM. The importance of proper experiment load case for the identification of coupled RFSs is revealed in (Link, Boeswald, Laborde, Weiland, & Calvi, 2011). Measurements of monocomponent harmonic oscillation only cover a single ellipse of the state space. To overcome this, sine wave excitation with consequently increased magnitude or stochastic excitation are suggested. In (Lenaerts, Kerschen, Golinval, Ruzzene, & Giorcelli, 2004) the RFSM is compared to a Wavelet Transform (WT) based parametric approach for the identification of nonlinear damping characteristics of a single mass oscillator.

A more empirical nonparametric representation of the system nonlinearities is called the Expanded Phase Space, where the three-dimensional trajectories, given by displacement ($\mathbf{z}$), velocity ($\mathbf{v}$) and acceleration ($\dot{\mathbf{v}}$), are considered (Volkova, 2013), (Volkova, 2011), (Volkova, 2010). These trajectories are strongly related to the RFSs and for the free oscillations of a 1DoF system they directly result in a mass-proportional RFS.

The main challenge of this method is obvious: it implies that time signals of both the complete state vector ($\mathbf{z}$ and $\mathbf{v}$) and all accelerations ($\dot{\mathbf{v}}$) are available. This can be achieved by either measuring all mentioned quantities or by generating the missing values through numerical integration or differentiation. While the latter can cause significant calculation errors, the former leads to complex, expensive or, in many cases, technically impossible measurement setups. Parallelly measuring all required signals can also cause relative phase shift between different quantities due to the necessary application of different sensor types. A restriction of the RFSM arises from equation (1.22), namely that only one equation per DoF is provided by the algorithm. From this it follows that only one RFS per degree of freedom is identifiable, i.e. the approach is only applicable to processes, where coupling between DoFs is restricted to the inertia matrix $\mathbf{M}$, q.v. equation (1.11). Furthermore, the method implies the a priori knowledge of $\mathbf{M}$. Detailed discussions on these difficulties and restrictions can be found in the previously cited literature on the RFSM.

Optimal Transformations

In (Voss, Rust, Horbelt, & Timmer, 2003) a nonparametric approach based on the Optimal Transformations technique is presented. This is a probabilistic method, where the RFCs and RFSs are determined in form of so called "optimal transformations" of the states and their derivatives, which are found by the maximisation of the correlation between the transformed states. These maximiser functions can be estimated in a nonparametric way using the Alternating Conditional Expectation algorithm (ACE). A weakness of this method arises in case of strong noise contamination of the measurements (Voss, Rust, Horbelt, & Timmer, 2003). This is mainly due to the fact that the nonmeasured state derivatives have to be generated numerically, which leads to the same difficulties as in case of the RFSM. The technique is implemented in a two-step algorithm in (Voss & Kurths, 1997) for time-delay systems, where multiple process time-delays are determined in addition to the nonlinear RFSs.

Method of State Dependant Parameters

The State Dependant Parameter (SDP) technique, which is based on optimal recursive estimation, is proposed in (Young, 2000). The method is derived as an extension of the technique of Time Variable Parameters (TVP). The restoring forces are described by time-varying linear coefficients and their time derivatives, called "instrumental variables", which are introduced as augmented states. Similar to the case of the HHT algorithm the time-varying parameter values are linked to the corresponding system state values to form RFS samples at every measurement time step. The approach addresses systems, where the dynamics can be given by an implicit algebraic equation of the form

$$\mathbf{y}(i) = \mathbf{h}\big(\mathbf{y}(i-1), \dots, \mathbf{y}(i-n_S), \mathbf{u}(i-1), \dots, \mathbf{u}(i-n_S)\big)\,, \qquad \textbf{(1.23)}$$

where the system is defined as a relationship between the current and previous values of the measurement. Therefore, the only internal states of the a priori model are the additionally introduced instrumental variables, which are described by a Generalised Random Walk (GRW) model. To avoid the necessity of fast parameter convergence, previously discussed in Section 1.4.1, an appropriate iterative sorting of the measurement time series is carried out prior to the identification. This solution crucially implies a system model in form of algebraic equations, i.e. (1.23). In (Young, 2000) the method is implemented for the identification of the RFCs of single input single output systems. The capability of the technique to identify coupled RFSs is mentioned in (Young & Garnier, 2006).

Summary of nonparametric methods

The main practical advantage of nonparametric approaches can be described as follows. On the one hand, if one prefers to determine models to gain insight to the underlying mechanisms in the system, the model selection is much easier in the case of nonparametric methods. For the selection, one has to answer questions regarding the nonlinearity, such as "Does it affect the elastic or the dissipative behaviour?", "Is it symmetric or asymmetric?" or "Is it smooth or non-smooth?". It is much easier to provide answers by looking at a point cloud representing the local RFS than by looking at the global system behaviour in the form of measurement signals. On the other hand, if one is satisfied with black-box models given by general mathematical functions, a relatively high number of parameters can be used to achieve quality results without computational problems because the global parameter estimation is now reduced to separate curve or surface fitting problems. As an extreme case one can completely resign the a posteriori characterisation, which results in Model on Demand (MOD) solutions (Ljung, 2010). During the fusion of the a priori model with the measurements only a few time-varying fitting parameters are introduced. The number of these parameters does not depend on the complexity of the nonlinearities involved in the RFSs that are to be estimated. This results in general a priori system models that are typically simpler than the ones needed in parametric algorithms

There is a risk of over fitting during the a posteriori characterisation of the RFSs due to the high resolution and possible noisiness of the RFS point clouds representing them. This can be overcome by applying proper probabilistic fitting and smoothing algorithms, which is discussed in Section 2.3.

1.5 Objective of the thesis

In the previous sections several aspects of the identification of nonlinear systems have been discussed. It has been shown that nonparametric methods have some noticeable advantages compared to parametric ones especially if no a priori knowledge of the system nonlinearities is available. However, from Section 1.4.2 it becomes clear that there is no state of the art nonparametric approach that is both suitable for the identification problem formulated in Section 1.3 without restrictions and fulfils the requirements from Section 1.1 at the same time. The frequency domain techniques require specific kinds of excitation of the system during the experiments and have difficulties in case of multiple DoF systems. The SDP approach is not compatible with dynamic systems given by differential equations. The RFSM fits the problem formulation from Section 1.3, it can be deployed for arbitrary types of excitation, it can identify general coupled RFSs and it is computationally simple at the same time, which makes it an extremely powerful technique. Unfortunately, the fact that it requires the time series of both the state vector and its derivatives and that it offers no possibility for the additional identification of the inertia matrix, represents a major drawback.

Using master-slave synchronisation between the virtual model and the real system for parametric identification of nonlinearities (in form of an augmented nonlinear Kalman Filter) is a well known technique with several published implementations (q.v. Section 1.4.1). However, the potential of this method for nonparametric identification, to the author's knowledge, has not yet been investigated in the literature. The current study is dedicated to explore this possibility with the objective of establishing an identification framework that exhibits all the advantageous properties of the RFSM but does not suffer from its drawbacks described above.

1.6 Kalman Filter

The current thesis presents a synchronisation-based nonparametric identification framework. It is based on the theory of nonlinear Kalman Filters, which is briefly summarised in the current section. It begins with the general theory of observers in Subsection 1.6.1, which is followed by a short overview of nonlinear probabilistic transformations in Subsection 1.6.2. These are crucial elements of nonlinear Kalman Filters that are briefly presented in Subsection 1.6.3.

1.6.1 Observers and synchronisation

The observation of dynamic systems is an important field of control theory. Without knowing the initial state, an observer is able to reconstruct the time history of a dynamic system's state vector based on the given system input $\mathbf{u}$ and the measured output $\mathbf{y}$. An observer realises a master-slave synchronisation using the system model given by (1.1) as the slave and the measured real system as the master. Such observers are essential parts of model predictive control and synchronisation algorithms of mechatronic systems (Nijmeijer & Angeles, 2003). The crucial relationship between observers and synchronisation is discussed in (Nijmeijer & Mareels, 1997). In (Santoboni, Pogromsky, & Nijmeijer, 2003) similar theory is presented for partially observable systems. An observer is generated through the modification of the process equation (1.1) by the introduction of the error between the real measurement signal and its estimated value given by (1.2). This error is called the synchronisation residual. The most convenient way of introducing this residual is using an additive linear correction term, which leads to a nonlinear observer of the form

$$\dot{\hat{\mathbf{x}}} = \mathbf{a}(\hat{\mathbf{x}}, \mathbf{u}) + \mathbf{K}_{n \times n_y}(\mathbf{y} - \hat{\mathbf{y}}), \tag{1.24}$$

$$\hat{\mathbf{y}} = \mathbf{h}(\hat{\mathbf{x}}, \mathbf{u}), \tag{1.25}$$

where $\mathbf{K}$ is the synchronisation gain matrix, and the "hat" symbol denotes estimates of the specific values. The gain matrix has to be chosen as a function of time in a way that the state estimation error converges to zero over time, i.e.

$$\mathbf{K}(t) \ni \lim_{t \to \infty}\left(\mathbf{x}(t) - \hat{\mathbf{x}}(t)\right) = \mathbf{0}\,. \tag{1.26}$$

The existence of a time sequence of $\mathbf{K}$ that fulfils (1.26) implies the observability of the system. This is ensured by the observability criterion, which can be formulated as a criterion for the rank of the observability matrix $\mathbf{O}$ (Hedrick & Girard, 2013), (Nijmeijer & Mareels, 1997):

$$\mathbf{O}_{\mathbf{x}_0,\mathbf{u}_0} := \left[\left[\frac{\partial \mathrm{L}^0_{\mathbf{a}}(h_1)}{\partial \mathbf{x}}, \dots, \frac{\partial \mathrm{L}^0_{\mathbf{a}}\left(h_{n_y}\right)}{\partial \mathbf{x}}\right], \dots, \left[\frac{\partial \mathrm{L}^{n-1}_{\mathbf{a}}(h_1)}{\partial \mathbf{x}}, \dots, \frac{\partial \mathrm{L}^{n-1}_{\mathbf{a}}\left(h_{n_y}\right)}{\partial \mathbf{x}}\right]\right], \tag{1.27}$$

where $\mathrm{L}^i_{\mathbf{a}}$ denotes the Lie derivative of order i with respect to the process vector function $\mathbf{a}$ (q.v. Appendix A), and h_j denotes the jth element of the measurement vector function $\mathbf{h}$. If $\mathbf{O}$ has a full rank n, then the system given by (1.1) and (1.2) is locally observable at $\{\mathbf{x}_0, \mathbf{u}_0\}$. According to (Hedrick & Girard, 2013) a physical explanation of this criterion is given by the Taylor series

$$\begin{bmatrix} \mathbf{y} \\ \dot{\mathbf{y}} \\ \vdots \\ \mathbf{y}^{(n-1)} \end{bmatrix}_{\mathbf{x},\mathbf{u}_0} = \begin{bmatrix} \mathbf{y} \\ \dot{\mathbf{y}} \\ \vdots \\ \mathbf{y}^{(n-1)} \end{bmatrix}_{\mathbf{x}_0,\mathbf{u}_0} + \mathbf{O}^{\mathrm{T}}_{\mathbf{x}_0,\mathbf{u}_0}(\mathbf{x} - \mathbf{x}_0) + \mathbf{e}_{\mathrm{HOT}}\,, \tag{1.28}$$

where $\mathbf{e}_{\mathrm{HOT}}$ refers to the error due to non-zero higher order terms. Therefore, the observability criterion ensures that (1.28) can approximately be solved for $\mathbf{x}$ in a local neighbourhood of $\mathbf{x}_0$. Practically speaking this means that $\mathbf{x}$ can be extracted from the time history of $\mathbf{y}$. Notice that according to (1.28) the estimation of $\mathbf{x}$ is possible using less measurement signals than its dimension, i.e. for

$n>n_y$, because an observer takes advantage of the expected dynamic behaviour of the system given by (1.1). In case of linear time-invariant (LTI) systems **O** is a constant matrix. Hence, for such systems the local criterion directly ensures global observability.

1.6.2 Probabilistic transformation of random variables

Since real life's measured signals are inevitably corrupted by noise (q.v. Section 2.2.1), the probabilistic variation of the injected measurement **y** is carried over to the estimated state variables and the estimated measurements through the observer equations (1.24) and (1.25) respectively. Hence, in order to be able to calculate the optimal time sequence of **K** that takes into account the random nature of the observer, the theory of transforming the mean and the variance of random variables has to be discussed first. The problem to be solved is the calculation of the mean vector **y** and covariance matrix $\mathbf{P_Y}$ of a random vector **Y** based on the known mean vector **x** and covariance matrix $\mathbf{P_X}$ of the random vector **X**, where **Y** is given by a general nonlinear transformation as $\mathbf{Y}=\mathbf{h}(\mathbf{X})$. (Without the loss of generality the chosen notation corresponds to the measurement equation (1.25).) For the sake of convenience a compact notation is proposed in the current thesis that represents a general probabilistic transformation, denoted by Φ, which gives an approximate solution to the above defined problem:

$$[\hat{\mathbf{y}} \quad \hat{\mathbf{P}}_{\mathbf{Y}} \quad \hat{\mathbf{P}}_{\mathbf{XY}}] = \Phi_{\mathbf{h}}^{k}(\mathbf{x}, \mathbf{P}_{\mathbf{X}}). \tag{1.29}$$

The additional result $\mathbf{P_{XY}}$ is the cross-covariance matrix that describes the stochastic correlation between **X** and **Y**. The superscript k indicates the type of the algorithm that is used to carry out the above transformation. A great number of such algorithms have been developed in the past sixty years. For a general function **h** all of them can only provide an estimate of the exact solution, which is indicated by the "hat" symbol in the notation of the resulting quantities. In the following the three most widespread algorithms are briefly summarised, two of which play a major role in the current study.

The most intuitive and general algorithm is the Monte Carlo Transformation, denoted by k=MCT. In this technique a finite number n_{MC} of samples, denoted by $\mathcal{X}_j$, are generated randomly according to $\mathbf{x}$ and $\mathbf{P}_{\mathrm{X}}$. These samples are then transformed into samples of $\mathbf{Y}$, denoted by $\mathcal{Y}_j$, using (1.30). Given these samples the equations (1.31), (1.32) and (1.33) are used to estimate the quantities of interest (Meyer, 2003).

$$\mathcal{Y}_j = \mathbf{h}(\mathcal{X}_j) \tag{1.30}$$

$$\hat{\mathbf{y}} = \frac{1}{n_{\mathrm{MC}}} \sum_{j=1}^{n_{\mathrm{MC}}} \mathcal{Y}_j \tag{1.31}$$

$$\widehat{\mathbf{P}}_{\mathrm{Y}} = \frac{1}{n_{\mathrm{MC}} - 1} \sum_{j=1}^{n_{\mathrm{MC}}} (\mathcal{Y}_j - \hat{\mathbf{y}})(\mathcal{Y}_j - \hat{\mathbf{y}})^{\mathrm{T}} \tag{1.32}$$

$$\widehat{\mathbf{P}}_{\mathrm{XY}} = \frac{1}{n_{\mathrm{MC}} - 1} \sum_{j=1}^{n_{\mathrm{MC}}} (\mathcal{X}_j - \mathbf{x})(\mathcal{Y}_j - \hat{\mathbf{y}})^{\mathrm{T}} \tag{1.33}$$

The MCT is a universal algorithm. The samples of $\mathbf{X}$ can be generated from arbitrary probability distributions, not only from those that can be approximately described by $\mathbf{x}$ and $\mathbf{P}_{\mathrm{X}}$ (i.e. approximately Gaussian distributions). As n_{MC} increases, the estimates converge to the exact solutions for any kind of $\mathbf{h}$, even for those that represent complex black-box algorithms. The only drawback is the slow statistical convergence that requires a large number of samples to achieve good result quality, which makes this approach computationally expensive.

In contrast to the MCT the Linearised Transformation, denoted by k=LinT, is an extremely simple algorithm. Using the equations (1.34), (1.35), (1.36) and (1.37) it performs a direct transformation without generating random samples (Simon, 2006).

$$\mathbf{H} = \left.\frac{\partial \mathbf{h}}{\partial \mathbf{X}}\right|_{\mathbf{x}} \tag{1.34}$$

$$\hat{\mathbf{y}} = \mathbf{h}(\mathbf{x}) \tag{1.35}$$

$$\widehat{\mathbf{P}}_{\mathbf{Y}} = \mathbf{H}\mathbf{P}_{\mathbf{X}}\mathbf{H}^{\mathrm{T}} \tag{1.36}$$

$$\widehat{\mathbf{P}}_{\mathbf{XY}} = \mathbf{P}_{\mathbf{X}}\mathbf{H}^{\mathrm{T}} \tag{1.37}$$

If **h** is a linear function and the distribution of **X** is Gaussian, then this approach gives the exact solution to the transformation. The less these two assumptions hold, the less accurate the technique becomes, therefore, it is only suitable for weak nonlinearities. Supposing that the Jacobian matrix of **h**, denoted by **H**, is easy to calculate (e.g. can be given analytically instead of being approximated via finite differences), then the LinT is computationally very cheap, since the function **h** is only calculated once in contrast to the MCT, where it is calculated n_{MC} times.

The Unscented Transformation (Julier & Uhlmann, 1997), denoted by k=UT, is a relatively new approach that managed to fill in the gap between the LinT and the MCT. It is suitable for much stronger nonlinearities than the LinT. Similar to the MCT it is derivative free (computation of **H** is not needed) and is also based on samples of the distribution of **X**. However, these samples, the so called "sigma points", are given by the deterministic formula (1.38) instead of being generated randomly, which enables a much faster statistical convergence. The number of sigma points is fixed to $n_{UT}=2n+1$, where n is the dimension of **X**. The rest of the algorithm consists of the transformation of the sigma points (1.39) and the analysis of the sampled statistics via (1.40), (1.41) and (1.42) similar to the MCT. The coefficients $p_{i.j}^{\mathrm{UT}}$ are essential tuning parameters. Their calculation is summarised in Appendix D.

$$[\mathcal{X}_1 \quad \mathcal{X}_2 \quad \cdots \quad \mathcal{X}_{n_{\mathrm{UT}}}] := \left[\mathbf{x} \quad \mathbf{x} - p_1^{\mathrm{UT}}\sqrt{\mathbf{P_X}} \quad \mathbf{x} + p_1^{\mathrm{UT}}\sqrt{\mathbf{P_X}}\right] \tag{1.38}$$

$$\mathcal{Y}_j = \mathbf{h}(\mathcal{X}_j) \tag{1.39}$$

$$\hat{\mathbf{y}} = \sum_{j=1}^{n_{\mathrm{UT}}} p_{2.j}^{\mathrm{UT}} \cdot \mathcal{Y}_j \tag{1.40}$$

$$\hat{\mathbf{P}}_{\mathbf{Y}} = \sum_{j=1}^{n_{\mathrm{UT}}} p_{3.j}^{\mathrm{UT}} (\mathcal{Y}_j - \hat{\mathbf{y}})(\mathcal{Y}_j - \hat{\mathbf{y}})^{\mathrm{T}} \tag{1.41}$$

$$\hat{\mathbf{P}}_{\mathbf{XY}} = \sum_{j=1}^{n_{\mathrm{UT}}} p_{3.j}^{\mathrm{UT}} (\mathcal{X}_j - \mathbf{x})(\mathcal{Y}_j - \hat{\mathbf{y}})^{\mathrm{T}} \tag{1.42}$$

1.6.3 Kalman Filters for nonlinear systems

If one has to address measurements corrupted by significant noise, the algorithm of choice for the calculation of the synchronisation gain **K** is the Kalman Filter (KF). This observer takes into account the uncertainty in the measurement equation (1.2) characterised by the measurement covariance matrix of size $n_y \times n_y$, denoted by **R**, as well as the uncertainty of the process equation (1.1) given by the discrete time process covariance matrix **Q**, which is of size $n \times n$. This turns the observer into a stochastic estimator that estimates the state and measurement vectors as probabilistic quantities **X** and **Y**. These are described by their mean values **x** and **y**, their covariance matrices $\mathbf{P_X}$ and $\mathbf{P_Y}$ as well as their cross-covariance matrix $\mathbf{P_{XY}}$ given as a joint distribution in the compact form

$$\begin{bmatrix} \mathbf{X} \\ \mathbf{Y} \end{bmatrix} \sim \left(\begin{bmatrix} \mathbf{x} \\ \mathbf{y} \end{bmatrix}, \begin{bmatrix} \mathbf{P}_\mathrm{X} & \mathbf{P}_\mathrm{XY} \\ \mathbf{P}_\mathrm{XY}^\mathrm{T} & \mathbf{P}_\mathrm{Y} \end{bmatrix} \right). \tag{1.43}$$

If finding the most probable time sequence of $\mathbf{x}$, i.e. the minimisation of $\mathrm{tr}(\mathbf{P}_\mathrm{X})$, is chosen as objective, then the Kalman gain, defined as

$$\mathbf{K}_i = \mathbf{P}_{\mathrm{XY},i} \mathbf{P}_{\mathrm{Y},i}^{-1} , \tag{1.44}$$

is the optimal solution to the estimation problem. The easiest way of deriving this elegantly simple formula is from a least squares point of view (Simon, 2006), however, a dynamic programming formulation is also possible (Cox, 1964). The so called continuous-time Kalman Filter, also known as the Kalman-Bucy filter (Kalman & Bucy, 1961), is of the form (1.24). However, due to the additive correction term, the observer equations can also be formulated in a two-step predictor-corrector form, which is of practical importance. In real life applications, the measurement signal is only available at discrete time points. To account for this, equation (1.24) is split into two recursively repeated steps. The first step is the time update (1.45) (also known as the prediction step), where the system state is integrated from the measurement time step *i*-1 to the time step *i* using equation (1.1) and an appropriate integration scheme. This results in the a priori state estimate, denoted by $\hat{\mathbf{x}}^-$. The second step is the measurement update (1.48) (also called the correction step), where the current synchronisation residual $(\mathbf{y}_i - \hat{\mathbf{y}}_i)$ is used to correct the predicted state, resulting in the a posteriori state estimate $\hat{\mathbf{x}}^+$, which is the final observer result for the *i*th time step. This approach is the so called discrete-time Kalman Filter (Kalman, 1960), which (using the notation proposed in Section 1.6.2) is given by the following compact recursive algorithm:

$$\begin{bmatrix} \hat{\mathbf{x}}_i^- & \widehat{\mathbf{P}}_{\mathrm{X},i}^- & \widehat{\mathbf{P}}_{\mathrm{XX},i}^- \end{bmatrix} = \Phi^k_{\hat{\mathbf{x}}_{i-1}^+ + \int_{t_{i-1}}^{t_i} \mathbf{a} dt} \left(\hat{\mathbf{x}}_{i-1}^+, \widehat{\mathbf{P}}_{\mathrm{X},i-1}^+ \right) + \begin{bmatrix} \mathbf{0} & \mathbf{Q} & \mathbf{0} \end{bmatrix} \tag{1.45}$$

$$\begin{bmatrix} \hat{\mathbf{y}}_i & \widehat{\mathbf{P}}_{\mathrm{Y},i} & \widehat{\mathbf{P}}_{\mathrm{XY},i} \end{bmatrix} = \Phi^k_\mathbf{h} \left(\hat{\mathbf{x}}_i^-, \widehat{\mathbf{P}}_{\mathrm{X},i}^- \right) + \begin{bmatrix} \mathbf{0} & \mathbf{R} & \mathbf{0} \end{bmatrix} \tag{1.46}$$

$$\mathbf{K}_i = \widehat{\mathbf{P}}_{\mathbf{XY},i} \widehat{\mathbf{P}}_{\mathbf{Y},i}^{-1} \tag{1.47}$$

$$\left[\hat{\mathbf{x}}_i^{+} \quad \widehat{\mathbf{P}}_{\mathbf{X},i}^{+}\right] = \left[\hat{\mathbf{x}}_i^{-} + \mathbf{K}_i(\mathbf{y}_i - \hat{\mathbf{y}}_i) \quad \widehat{\mathbf{P}}_{\mathbf{X},i}^{-} - \mathbf{K}_i \widehat{\mathbf{P}}_{\mathbf{Y},i} \mathbf{K}_i^{\mathrm{T}}\right] \tag{1.48}$$

In order to start the algorithm the initial a posteriori distribution of **X** has to be defined via $\hat{\mathbf{x}}_0^{+}$ and $\widehat{\mathbf{P}}_{\mathbf{X},0}^{+}$. Furthermore, **Q** and **R** need to be given according to the expected uncertainties. The proper choice of these four user defined quantities is discussed in Chapter 2.2.4. The recursion is continued until the last measurement time step is processed. Although the main result of the KF is the estimated mean of the state vector $\hat{\mathbf{x}}_i^{+}$, the additional covariance output $\widehat{\mathbf{P}}_{\mathbf{X},i}^{+}$ can be used as well as an indicator of the expectable error and further statistical properties of the estimated state (q.v. Section 2.3.1).

If the system functions (**a** and **h**) are nonlinear, the accuracy of the KF depends on the type *k* of the probabilistic transformation Φ, for which only approximate solutions exist for general nonlinearities (q.v. Section 1.6.2). Several different solutions to this problem have been developed over the past five decades since the Kalman Filter was proposed. In (Gelb, Kasper, Nash, Price, & Sutherland, 2001), (Simon, 2006) and (Hartikainen, Solin, & Särkkä, 2011) different selections of these algorithms are derived, discussed and compared and in (Moreno & Pigazo, 2009) many advanced application studies and special KF formulations are reported. Due to their simple formulation combined with good estimation accuracy, two specific algorithms have gained notable attention in the literature and in practical applications. The first one is the "classical" Extended Kalman Filter (EKF) that uses *k*=LinT. It is, therefore, only suitable for weak nonlinearities and approximately Gaussian distributions (q.v. Section 1.6.2). The second one is a relatively new technique called the Unscented Kalman Filter (UKF) (Julier & Uhlmann, 1997), which is based on *k*=UT. Hence, it can better account for strong system nonlinearities than the EKF, because they do not have to be linearised (q.v. Section 1.6.2). It can be formulated in an enhanced square-root form, called the Square-Root Unscented Kalman Filter (SR-UKF). The algorithm remains the same from the analytical point-of-view, however, the SR-UKF increases the numerical stability of the UKF and slightly reduces its computational time. For an exhaustive yet compact description of the algorithm the reader is referred to (van der Merwe & Wan, 2001). Different types of Sigma-Point Kalman Filters (SPKF) are presented in (van der Merwe & Wan, 2004) as general formulations of the UKF. If the nonlinearities become so severe that the accuracy of the UKF is not sufficient anymore, the Particle Filter (PF) can be deployed as a universal solution (Simon, 2006), (Chatzi & Smyth, 2009). This technique basically uses k=MCT (with some additional algorithm refinements), which on the one hand can handle arbitrary nonlinearities, but on the other hand leads to an enormous increase of computational costs (q.v. Section 1.6.2).

A special offline extension of the KF, the so called Kalman Smoother (KS), can be used to enhance result quality by involving not only past and present but also future measurement samples in the estimation algorithm for a given time step (Hartikainen, Solin, & Särkkä, 2011), (Simon, 2006), (Nicklas, 1989). The Fixed-Interval RTS Smoother (named after Rauch, Tung and Striebel) is a computationally efficient global form of the KS, which involves the whole available measurement time interval to generate an optimal estimate of the state vector's complete time series. It is defined as a backward recursion over time that is carried out in addition to the standard KF run based on its results. According to (Hartikainen, Solin, & Särkkä, 2011) its algorithm is given as

$$\mathbf{K}_i^{\mathrm{S}} = \widehat{\mathbf{P}}_{\mathbf{XX},i+1}^{-}\left(\widehat{\mathbf{P}}_{\mathbf{X},i+1}^{-}\right)^{-1} \tag{1.49}$$

$$\left[\hat{\mathbf{x}}_i^{\mathrm{S}} \quad \widehat{\mathbf{P}}_{\mathbf{X},i}^{\mathrm{S}}\right] = \left[\hat{\mathbf{x}}_i^{+} + \mathbf{K}_i^{\mathrm{S}}\left(\hat{\mathbf{x}}_{i+1}^{\mathrm{S}} - \hat{\mathbf{x}}_{i+1}^{-}\right) \quad \widehat{\mathbf{P}}_{\mathbf{X},i}^{+} + \mathbf{K}_i^{\mathrm{S}}\left(\widehat{\mathbf{P}}_{\mathbf{X},i+1}^{\mathrm{S}} - \widehat{\mathbf{P}}_{\mathbf{X},i+1}^{-}\right)\left(\mathbf{K}_i^{\mathrm{S}}\right)^{\mathrm{T}}\right] \tag{1.50}$$

The so called smoother gain is thereby denoted by $\mathbf{K}^S$, whereas the smoothed estimates of the state vector's mean and covariance are denoted by $\hat{\mathbf{x}}_i^S$ and $\hat{\mathbf{P}}_{\mathbf{X},i}^S$ respectively.

One difficulty with the application of KFs is their sensitivity to model accuracy. The rigorous solution to this problem is the introduction of every uncertain model parameter as state variable, which has been implemented in the current thesis (q.v. Section 2.1.2). Yet, a lot of research has been invested in robustified KF algorithms that are designed to handle model errors automatically. Such an adaptive KF is discussed in (Pulido Herrera & Kaufmann, 2010) that adjusts $\mathbf{Q}$ and $\mathbf{R}$ based on the statistics of the synchronisation error. The Schmidt-Kalman Filter, which can compensate particular model uncertainties, is presented in (McBurney, 1990). A more sophisticated way of robust filtering is the H_∞ approach (Simon, 2001), (Simon, 2006). These algorithms offer an easy way of compensating model errors, hence, they are very useful in monitoring and target tracking applications. However, these techniques do not attempt to identify the model's deviation from the real system, therefore, they are not suitable for the present study.

1.7 Curve and surface fitting

As discussed in Section 1.4.2, nonparametric identification approaches generate a point cloud of samples for each RFS while the measurement signals are processed. In most of these techniques, such as the one that is proposed in the current thesis, the noisiness of the measured signals is carried over to the RFS samples. Therefore, the a posteriori application of curve and surface fitting approaches is required to achieve noisefree RFCs and RFSs. In order to clarify the advantages and challenges of the fitting algorithms, which are implemented in the present study (q.v. Section 2.3.2 and 2.3.3), a brief overview of existing approaches is given in the current section.

To illustrate the problem let us consider the case of fitting an elastic RFC, i.e. $c_E(z)$ according to (1.9), using noisy samples given by the coordinate pairs $\{z_i, f_{E,i}\}$ (q.v. Section 1.4.2). In general the sample coordinates are corrupted by correlated noise, which is described by the noise covariance matrix, denoted by $\mathbf{P}_{cE}$. The curve fitting problem can be formulated as the minimisation of the Weighted Total Squares error e^{WTS} defined as

$$\mathbf{e}_{cE,i} = \begin{bmatrix} z_i \\ f_{E,i} \end{bmatrix} - \begin{bmatrix} \hat{c}_E^{-1}(f_{E,i}) \\ \hat{c}_E(z_i) \end{bmatrix}, \tag{1.51}$$

$$e_{cE}^{\mathrm{WTS}} = \sum_{i=1}^{n_s} \mathbf{e}_{cE,i}{}^{\mathrm{T}} \mathbf{P}_{cE,i}^{-1} \mathbf{e}_{cE,i}\,, \tag{1.52}$$

where $\hat{c}_E$ and $\hat{c}_E^{-1}$ denote the estimate of the true RFC and its inverse function respectively. The function $\hat{c}_E$ is an optimal estimate in Weighted Total Least Squares (WTLS) sense if it is the minimiser of e_{cE}^{WTS}.

The literature offers different solutions to the WTLS estimation problem. It is often referred to as the Errors-In-Variables (EIV) problem, which clearly differentiates it from the ordinary Least Squares (LS), where the abscissas (independent variables) of the observations are assumed to be noisefree. The first group of these methods are parametric. They imply an assumption of the analytical form of the function of interest, whose parameters are estimated by the procedure. A solution of this kind in a standard TLS form based on Singular Value Decomposition (SVD) is proposed in (Golub & Van Loan, 1980), where abscissas and ordinates are equally weighted. (Markovsky & Van Huffel, 2007) presents a hierarchical comparison of parametric WTLS methods regarding the special cases of weighting that they take into account. The general WTLS problem however has no direct analytical solution. An iterative technique based on a modified Recursive Weighted LS (RWLS) formulation is proposed in (Amiri-Simkooei & Jazaeri, 2012), where the abscissas and the ordinates of the observations are arbitrary weighted, they cross-covariance (i.e. they correlation) is however not taken into account. The second group of fitting methods carry out a nonparametric regression, i.e. they seek an optimally smoothed version of the observed dataset without the assumption of a global analytical form of the characteristic. This smoothed dataset itself forms the solution to the fitting problem, i.e. $\hat{c}_E$ in the current case, as a piecewise defined function that is given in form of a lookup table. An overview of such regression approaches is given in (Ljung, 2010). One of them is referred to as Local Polynomial Regression (LPR), where the smoothed version of each observation is generated via a polynomial regression involving a weighted group of observations in its neighbourhood. A local LS formulation of this technique is discussed and compared with smoothing splines in (Fox, 2002) for 2D curves and 3D surfaces. A TLS extension of the approach, called the Improved Moving Least Squares (IMLS) method, is proposed in (Zhang, et al., 2013), where however only the special case of uncorrelated error is considered.

1.8 Thesis outline

Chapter 2 presents the Kalman Filter-based nonparametric method in detail that represents an alternative solution to the identification problem defined in Section 1.3. This is followed by virtual identification examples of one and three DoF nonlinear mechanical systems in Chapter 3 to demonstrate the effectiveness of the proposed technique. Based on these examples several implementation properties of the approach are investigated and advantages as well as challenges in comparison to state of the art methods are discussed.

In Chapter 4 the combination of the proposed technique with parametric approaches is investigated in order to carry out a full system identification that includes the a priori defined uncertain parameters of the system as well. These parameters are assumed to be known in Chapter 2 and 3.

Finally in Chapter 5 real-life implementation is presented on a 1DoF and a 3DoF mechanical system to prove the reliability of the algorithm under realistic conditions. The main outcomes of the thesis and open questions are then summarised in Chapter 6.

2 Nonparametric identification of RFSs

In the following a synchronisation-based nonparametric identification method is presented that allows the reconstruction of the noisefree RFSs according to the problem formulation in Section 1.3. In this chapter we assume that the inertia matrix **M** and the input matrix **B** are a priori known. This assumption is relaxed later in Chapter 4. The main idea of the current approach was proposed in (Kenderi & Fidlin, 2014). This chapter presents the final stage of the complete workflow including the extension towards coupled RFSs, detailed mathematical descriptions and several algorithm refinements. The workflow consists of two major steps:

- The first step is the fusion of the measurement signals with the a priori system model. It is carried out by means of the Kalman Filter based synchronisation of the virtual model to the real system. This results in recursive optimal estimates of properly chosen instrumental variables (augmented states) and their variances, which are directly related to local RFS samples. The establishment of the particular nonlinear Kalman Filter is presented Section 2.1 and 2.2.
- The second step begins with the probabilistic transformation of the estimated time series of the augmented states into RFS samples. This is followed by optimal nonparametric fitting, which results in noisefree RFCs and RFSs. These elements of the approach are derived in Section 2.3.

At the end of the chapter, in Section 2.4, all steps of the presented algorithm are summarised in a compact form.

2.1 Ensuring observability

2.1.1 Practical aspects

System identification relies on the close co-operation between virtual modelling and experiment design. Therefore, it is inevitable to take some practical aspects of vibration testing into account. Displacement, velocity and acceleration (or their corresponding rotational equivalents) are the commonly measured kinematic quantities in praxis for mechanical structures (Chaurasiya, 2012). In Section 1.4 it is discussed that identification methods differ in the necessary measurement setup. Some methods require all measurement types at the same time. Other techniques have the advantage that they allow the use of only one of the mentioned quantities. Many of these techniques (e.g. frequency domain approaches) exhibit the nontrivial property that they are functional with one arbitrary measurement type. This chapter presents a synchronisation-based nonparametric approach that provides the observability of the RFSs using only one arbitrary measured kinematic quantity, i.e. it exhibits the mentioned advantageous property as well. In order to point out, why this is so important, some basic aspects of choosing the proper measurement quantity for a vibration experiment are discussed in the following. An overview of these aspects can be found e.g. in (Harris & Piersol, 2002). They can be briefly summarised as follows:

- *Displacement or strain measurement* is suitable for low frequency range, where the magnitude of velocity and acceleration is typically low. Such relatively slow motions occur e.g. in civil engineering structures. This quantity should be chosen if exact positions are particularly important, e.g. possible collision of system components.
- *Velocity measurement* is ideal for mid frequency range. It typically results in a more uniform spectrum than displacement or acceleration.

This quantity is preferred for acoustic investigations, since sound pressure is proportional to the normal velocity of the vibrating surface.

- *Acceleration measurement* is the quantity of choice for high frequency range, where the magnitude of displacement or velocity signals is too small to be measured accurately. This is the case for most oscillation phenomena in mechanical engineering, such as noise issues. Typically this measurement type is preferred if a transducer of small size and small mass is required, and contactless methods are not deployable, since accelerometers are mostly smaller than other transducers. Additionally, typical accelerometers do not require a reference point and therefore usually lead to simple measurement setups.
- *Force measurement* differs from the previous measurement types in that it utilises a kinetic quantity. Such sensors have to be placed into the force flow of the system. This means that in case of measuring response forces inside a structure the transducer itself becomes a part of the mechanical system and can significantly alter its behaviour. Due to this difficulty, such sensors are mainly used to measure the external excitation force $\mathbf{u}$ acting on the structure during the experiment.

It should be mentioned that in the vibration trajectories of strongly nonlinear systems, which are of particular interest in the present study, higher harmonics of the main oscillation frequency are present. Due to this, transducers for higher frequency domains can be advantageous for the identification of such systems even if the dominant vibration frequency is rather low. Therefore, accelerometers have specific importance in case of nonlinear mechanical structures. However, on the whole it is clear that a flexible identification method should accommodate all possible kinematic measurement quantities.

Another important practical aspect of system identification is the unavoidable presence of noise in measured signals. This natural imperfection of measurements has a major influence on the way the estimated system states are to be handled in an observer algorithm. This topic is discussed later in Section 2.2.1, since it is not directly related to observability.

2.1.2 The a priori model of choice

As already proposed in the previous sections the aim of the present work is to establish an observer-based approach that solves the identification problem formulated in Section 1.3 in a nonparametric form. To achieve this according to Section 1.6.1 the process vector function **a** in equation (1.24) has to be brought into a form that provides the observability of RFS coordinate triplets $\{z,v,f\}_i$ for all kind of kinematic measurement types. In the following, this observability problem is investigated for the 1DoF case. Afterwards, the results are generalised for multiple DoF systems. That means, a system of the form

$$\begin{bmatrix} \dot{z} \\ \dot{v} \end{bmatrix} = \begin{bmatrix} v \\ -m^{-1}c(z,v) + bu \end{bmatrix}, \tag{2.1}$$

$$y = h(w,u), \;\; w \in \{z,v,\dot{v}\}, \tag{2.2}$$

is considered first, where z and v are the deformation and rate of deformation coordinates of the system's RFS respectively. The excitation in the form of force or acceleration is denoted by u, and b is the input coefficient. The mass of the 1DoF oscillator is denoted by m, and is assumed to be a priori known throughout this chapter. A first idea could be to make some assumptions about the form of $c(z,v)$ and hope that using (1.24) the system states will synchronise to their real values. In that case the missing third coordinate of the RFS sample triplets (i.e. f) could be calculated using the RFSM method from Section 1.4.2. Unfortunately this would not be the case, which becomes clear if we take a look at the observer equations (1.24) and (1.25) in detail. Notice that the system describing functions (**a** and **h**) appear in the equations without the "hat" symbol, i.e. the basic concept of the observer implies that these functions are assumed correctly. Otherwise the state vector can not converge to its real values perfectly. In case of uncertain (unknown) parameters in the system functions, the solution to this problem is the so-called state augmentation technique. This introduces all uncertain parameters as additional state variables to achieve a correct model assumption and to let the observer estimate all the uncertain values. Applying this approach, the augmented state vector, denoted by $\mathbf{x}_a$, will be of size n_a, where the number of unknown system parameters equals n_a-n. This is a well known technique for parametric system identification and for the estimation of time-varying system parameters with many successful implementations in the literature (q.v. Section 1.4.1). Based on this idea the simplest solution is to introduce the restoring force f as a state variable which leads to the process equation

$$\begin{bmatrix} \dot{z} \\ \dot{v} \\ \dot{f} \end{bmatrix} = \begin{bmatrix} v \\ -m^{-1}f + bu \\ 0 \end{bmatrix}, \tag{2.3}$$

where the augmented state vector directly represents the RFS sample triplets. Since no information about the expected time-variation of the new state variable is available a priori, it is modelled as a constant state. However, via proper tuning of the observer, which is discussed later in Section 2.2.4, the estimates of such states can be updated fast enough to track their oscillations. But to ensure convergence according to (1.27) the observability has to be verified first by taking a look at the observability matrices of the system for the three possible measurement types: displacement (z), velocity (v) and acceleration ($\dot{v}$), which are of the following form:

$$\mathbf{O}|_{y=z} = \begin{bmatrix} 1 & 0 & 0 \\ 0 & 1 & 0 \\ 0 & 0 & -\dfrac{1}{m} \end{bmatrix}, \mathbf{O}|_{y=v} = \begin{bmatrix} 0 & 0 & 0 \\ 1 & 0 & 0 \\ 0 & -\dfrac{1}{m} & 0 \end{bmatrix}, \mathbf{O}|_{y=\dot{v}} = \begin{bmatrix} 0 & 0 & 0 \\ 0 & 0 & 0 \\ -\dfrac{1}{m} & 0 & 0 \end{bmatrix}. \tag{2.4}$$

Recall from (1.28) that the *j*th row of **O** represents a linearised relation between the *j*th state variable and the measurement. This means that according to (2.4) the restoring force f remains observable for all measurement cases. However, for $y=v$ the observability of z is lost, and for $y=\dot{v}$ both z and v become unobservable. Therefore, the following conclusion can be drawn about directly introducing the restoring force as state variable:

- It provides observability of RFS samples for the special case of displacement measurements.
- It enables the observation of the time history of f for all measurement types, which is useful if one is interested in the internal forces of a system without directly relating them to phase plane coordinates.

However, according to the reasoning in Section 2.1.1 this is not a satisfactory result in the current case. Therefore, a more advantageous system model is needed that introduces no restrictions on the measured kinematic quantity. As a solution to the problem the following form of the process equation is proposed:

$$\begin{bmatrix} \dot{z} \\ \dot{v} \\ \dot{s} \\ \dot{d} \end{bmatrix} = \begin{bmatrix} v \\ -m^{-1}(sz + dv) + bu \\ 0 \\ 0 \end{bmatrix}, \qquad \textbf{(2.5)}$$

where s and d denote time-varying effective stiffness and effective damping coefficients respectively with a priori unknown time history, which we introduce as state variables to be estimated by the observer. Due to this modification the f coordinate of the RFS samples is not directly observed anymore. So called "instrumental variables" (Young, 2000) have been introduced instead that allow the a posteriori reconstruction of the actual quantity of interest, in this particular case given by (2.6). Notice that the modified formulation even allows the direct reconstruction of the separate elastic and dissipative restoring forces

(f_E and f_D respectively) in case of additive RFSs using the equations (2.7). This is of practical advantage, which is discussed in detail in Section 2.3.

$$f_i = s_i z_i + d_i v_i \tag{2.6}$$

$$f_{E,i} = s_i z_i \quad , \quad f_{D,i} = d_i v_i \tag{2.7}$$

In the following further properties of the defined model are discussed in detail, which starts with the analysis of the observability matrices. For the three measurement types the matrices consist of repeated columns. Therefore, for the sake of simplicity their main structure is given in equations (2.8), (2.9) and (2.10) for displacement, velocity and acceleration measurements respectively, while the analytical expressions of their column vectors $\mathbf{o}_j$ are listed separately in equations (2.11), (2.12) and (2.13). In contrast to (2.4) these matrices exhibit a coupled structure with more than one non-zero entries in their rows. This makes the investigation of their symbolic rank difficult. However, in the single measurement case the observability matrix is quadratic, hence, its non-zero determinant can be used as a proof of its full rank. Equations (2.14), (2.15) and (2.16) show the symbolic determinants of the matrices (2.8), (2.9) and (2.10) respectively.

$$\mathbf{O}|_{y=z} = [\mathbf{o}_1 \quad \mathbf{o}_2 \quad \mathbf{o}_3 \quad \mathbf{o}_4] \tag{2.8}$$

$$\mathbf{O}|_{y=v} = [\mathbf{o}_2 \quad \mathbf{o}_3 \quad \mathbf{o}_4 \quad \mathbf{o}_5] \tag{2.9}$$

$$\mathbf{O}|_{y=\dot{v}} = [\mathbf{o}_3 \quad \mathbf{o}_4 \quad \mathbf{o}_5 \quad \mathbf{o}_6] \tag{2.10}$$

$$\mathbf{o}_1 = \begin{bmatrix} 1 \\ 0 \\ 0 \\ 0 \end{bmatrix}, \mathbf{o}_2 = \begin{bmatrix} 0 \\ 1 \\ 0 \\ 0 \end{bmatrix}, \mathbf{o}_3 = \begin{bmatrix} -\frac{s}{m} \\ -\frac{d}{m} \\ -\frac{z}{m} \\ -\frac{v}{m} \end{bmatrix}, \mathbf{o}_4 = \begin{bmatrix} \frac{sd}{m^2} \\ \frac{d^2 - sm}{m^2} \\ \frac{dz - vm}{m^2} \\ \frac{2dv + sz - bum}{m^2} \end{bmatrix} \tag{2.11}$$

$$\mathbf{o}_5 = \begin{bmatrix} \dfrac{s^2m - sd^2}{m^3} \\ \dfrac{2smd - d^3}{m^3} \\ \dfrac{2dvm + 2msz - bum^2 - zd^2}{m^3} \\ \dfrac{2svm + 2dbum - 3d^2v - 2dsz}{m^3} \end{bmatrix} \tag{2.12}$$

$$\mathbf{o}_6 = \begin{bmatrix} \dfrac{sd^3 - 2s^2dm}{m^4} \\ \dfrac{s^2m^2 - 3sd^2m + d^4}{m^4} \\ \dfrac{2m^2sv + 2m^2dbu - 4mdsz - 3md^2v + d^3z}{m^4} \\ \dfrac{2sm^2bu - 2s^2zm - 6dsvm - 3bud^2m + 4d^3v + 3d^2sz}{m^4} \end{bmatrix} \tag{2.13}$$

$$\det\left(\mathbf{O}|_{y=z}\right) = -m^{-3}[sz^2 + mv^2 + dzv - zmbu] \tag{2.14}$$

$$\det\left(\mathbf{O}|_{y=v}\right) = -m^{-5}s\,[(mbu - sz)^2 + msv^2 + szdv - dvmbu] \tag{2.15}$$

$$\det\left(\mathbf{O}|_{y=\dot{v}}\right) = -m^{-7}s^3[(mbu - sz)^2 + msv^2 + szdv - dvmbu] \tag{2.16}$$

Due to the modification of the process model introduced in (2.5), the determinant is symbolically non-zero for all three measurement types, which makes observability available in all cases. However, because the process model has become nonlinear, the determinants describe local criteria. They are functions of variables and hence can become zero at certain variable combinations. The right-hand sides of the equations are written as products of a mass- and effective stiffness-dependent factor and an expression within square brackets. The first leads to the criterion that s is not allowed to be zero if velocity or acceleration is measured. The effect of zero effective stiffness becomes clear by taking a look at the first row vectors of the observability matrices, i.e. the first entries in the column vectors $\mathbf{o}_j$. Except for $\mathbf{o}_1$, which only appears in $\mathbf{O}$ for y=z, these entries become zero for s=0 leading to the loss of the displacement's observability. This means that for y=v and y=$\dot{v}$ the virtual model can only synchronise with the measured system as long as a symbolic connection between the derivative of the state vector and z exists in the process equation. On the other hand this is actually the reason why the introduced model allows observability for $y\epsilon\{v,\dot{v}\}$, because a symbolic relationship between the restoring force and z has been introduced.

Getting back to the determinants of the observability matrices, the expressions inside the square brackets still need to be discussed. In order to decide under which conditions these expressions can become zero, a few algebraic manipulations are necessary. Let us begin with the case y=z. From the second row of the process equation (2.5) the equality (2.17) can be derived.

$$mbu = m\dot{v} + dv + sz \tag{2.17}$$

Using (2.17) to replace mbu and carrying out some simplifications the symbolic determinant from (1.19) becomes

$$\det\left(\mathbf{O}|_{y=z}\right) = -m^{-3}[\, mv^2 - zm\dot{v}\,]\,. \tag{2.18}$$

For further analysis let us assume that the system's behaviour can be described locally as a harmonic oscillation with an instantaneous amplitude Λ and an instantaneous angular frequency ω, which leads to

$$z = \Lambda \cdot \sin(\omega t), \tag{2.19}$$

$$v = \omega \cdot \Lambda \cdot \cos(\omega t), \tag{2.20}$$

$$\dot{v} = -\omega^2 \cdot \Lambda \cdot \sin(\omega t). \tag{2.21}$$

This assumption is only reasonable for s>0. Therefore, from this point on the symbolic investigation of observability is restricted to the subclass of RFSs that have locally positive effective stiffness over the phase plane range of interest. According to the locally harmonic assumption (2.18) turns into

$$\det\left(\mathbf{O}|_{y=z}\right) = -m^{-3}[\, m\omega^2 \Lambda^2\{\sin^2(\omega t) + \cos^2(\omega t)\}\,]\,, \tag{2.22}$$

which is apparently never equal to zero as long as the system is in motion. Therefore, the system (including its RFS) is always observable for *y*=*z*. This actually satisfies a natural requirement, since the unmodified process model (2.3) already exhibited this property.

The other two measurement cases, i.e. $y\epsilon\{v, \dot{v}\}$, can be analysed together, since the expressions inside the square brackets in (2.15) and (2.16) are the same. To do so, we take advantage of the following two equality expressions. To get the equality (2.23) the square of the second row of the process equation (2.5) is taken and then rearranged. The equality (2.24) follows from the same equation after the multiplication of both sides by *dv* and subsequential rearrangement of its terms.

$$(mbu - sz)^2 = m^2\dot{v}^2 + 2dvmbu - 2szdv - d^2v^2 \tag{2.23}$$

$$dvmbu - szdv - d^2v^2 = dvm\dot{v} \tag{2.24}$$

After successive substitution of the right-hand sides of (2.23) and (2.24) for their left-hand sides in the expression inside the square bracket in the symbolic observability determinants for $y\epsilon\{v, \dot{v}\}$, it becomes

$$m^2\dot{v}^2 + msv^2 + dvm\dot{v}\,. \tag{2.25}$$

Applying again the locally harmonic system description using (2.20) and (2.21), the expression (2.25) turns into

$$m\omega^2\Lambda^2\{\, m\omega^2 \sin^2(\omega t) + s\cos^2(\omega t) - d\omega\cos(\omega t)\sin(\omega t)\,\}\,. \tag{2.26}$$

Only the part inside the curly bracket is discussed further, since the multipliers outside the bracket never become zero as long as the system is in motion. After deploying some trigonometric equalities (q.v. Appendix A) and carrying out some symbolic simplifications, the expression within the curly bracket becomes

$$\frac{1}{2}(s + m\omega^2) - \frac{1}{2}\sqrt{(s - m\omega^2)^2 + (d\omega)^2}\sin(2\omega t + \theta)\,. \qquad \textbf{(2.27)}$$

Since we consider the case of s>0, the expression (2.27) can not become zero as long as the positive constant term is greater than the amplitude of the harmonic term, i.e. as long as (2.28) holds, which after some algebraic manipulations reduces to the criterion (2.29).

$$\sqrt{s^2 - 2sm\omega^2 + m^2\omega^4 + d^2\omega^2} < s + m\omega^2 \qquad \textbf{(2.28)}$$

$$\frac{d^2}{4m} < s \qquad \textbf{(2.29)}$$

This means that for $y \epsilon \{v, \dot{v}\}$ the instantaneous effective stiffness is not only not allowed to be zero, but it has to be higher than a certain level, which is defined by (2.29). The formula can also be rearranged into a criterion for the instantaneous effective damping as

$$\frac{|d|}{2\sqrt{sm}} < 1\,. \qquad \textbf{(2.30)}$$

Notice that the left-hand side of (2.30) is the linearised damping ratio and its maximum allowed value is exactly the "critical damping" (Harris & Piersol, 2002), which means that in case of $y \epsilon \{v, \dot{v}\}$ observability is only provided if the locally linearised system can exhibit free oscillation.

Now that the form of the general a priori model has been established for the 1DoF case, it can be generalised for the multiple DoF system from Section 1.3, which leads to an observable (and hence by synchronisation identifiable) form of the *N*DoF process equation:

$$\begin{bmatrix} \dot{\mathbf{z}}_{N\times1} \\ \dot{\mathbf{v}}_{N\times1} \\ \dot{\mathbf{s}}_{N\times1} \\ \dot{\mathbf{d}}_{N\times1} \end{bmatrix} = \begin{bmatrix} \mathbf{v}_{N\times1} \\ -\mathbf{M}_{N\times N}^{-1}[\mathrm{diag}(\mathbf{s}_{N\times1})\mathbf{z}_{N\times1} + \mathrm{diag}(\mathbf{d}_{N\times1})\mathbf{v}_{N\times1}] + \mathbf{B}_{N\times n_u}\mathbf{u}_{n_u\times1} \\ \mathbf{0}_{N\times1} \\ \mathbf{0}_{N\times1} \end{bmatrix}. \quad \textbf{(2.31)}$$

The entries **s** and **d** are the vectors of the N time-varying effective stiffness and damping coefficients respectively, therefore, the augmented state vector, denoted by $\mathbf{x}_a$, is of size n_a=2n. Because observability is a property of the system and not of the process equation alone, the measurement equation also has to be defined in a way that allows the observation of the augmented states. Since the effective stiffness and damping coefficients are unknown states, the a priori model has no information about the oscillatory modes of the system. Due to this the process equation does not carry a clear connection between DoFs, hence, the necessary measurement setup can be formulated as

$$\mathbf{w}_{N\times1} \coloneqq [w_1, \dots, w_N]^{\mathrm{T}} \ni w_j \in \{z_j, v_j, \dot{v}_j\}, \quad \textbf{(2.32)}$$

$$\mathbf{y}_{N\times1} = \mathbf{h}(\mathbf{w}, \mathbf{u}) \ni \mathrm{rank}\left(\frac{\partial \mathbf{h}}{\partial \mathbf{w}}\right) = N, \quad \textbf{(2.33)}$$

where **w** is an N element vector consisting of w_j, which can arbitrarily be chosen to be the 0th, 1st or 2nd time derivative of the jth element of the displacement (deformation) vector **z**. The measurement vector **y** is of size n_y=N, which indicates that the necessary number of measurement signals that ensures the observability of the augmented state vector is equal to the number of DoFs in the mechanical structure. Additionally, the measurement equation has to be solvable for **w**, which is guaranteed by the rank criterion included in (2.33).

2.1.3 Discussion on the chosen model

Considering the aspects from Section 2.1.1 the a priori system model introduced in Section 2.1.2 is suitable for the nonparametric identification of RFSs. It has a very simple structure that provides observability using one measured arbitrary mechanical quantity per DoF. Compared to directly observing the restoring force (2.3) the introduced model (2.5) provides observability for velocity and acceleration measurements as well. The only price to be paid for this property is that the system equations become nonlinear functions of the augmented states. The introduced bilinear structure $c(z,v){:=}sz{+}dv$ however can be considered as a weak nonlinearity, which remains in the same form regardless of the complexity of the RFS that is to be identified. Despite of all these fruitful properties the following three minor drawbacks still remain:

(1) The loss of observability for velocity and acceleration measurements in case of effective stiffness values that does not satisfy (2.29) requires additional state constraints in the observer (q.v. Section 2.2.5), and causes complications for some rare types of RFSs, where the effective stiffness changes its sign at some points of the phase plane.

(2) In case of the identification of coupled RFSs two instrumental variables (s and d) are used to estimate one quantity of interest (f).

(3) The fact that s and d are assumed as constant states in the process equation requires fast convergence of the observer over time.

Since the present model choice seems rather intuitive, the question arises from natural suspicion, whether there is a better model formulation that can get rid of the mentioned difficulties. This topic is discussed briefly in the following.

(1) *Is it possible to provide observability for zero effective stiffness in case of velocity and acceleration measurements?*

The difficulty with $s{=}0$ is not a specific property of the current approach. It is a consequence of the fact that a state (in this case z) can not be directly observed from the measurement of its time derivatives (in this case v or $\dot{v}$). Therefore, it can be concluded that a symbolic connection between the time derivative of the state vector and z is always required for $y \in \{v, \dot{v}\}$ in order for z to be observable.

Since in the current problem formulation (q.v. Section 1.3) the RFSs carry the only connection between the internal forces and the displacements, it is not possible to find another form of the a priori model that would overcome this restriction, because the connection is always lost if the elastic restoring force becomes zero. It might be possible to find a model structure that would allow the effective stiffness to get nearer to zero than (2.29). However, numerical investigations (q.v. Chapter 3) show that this criterion can successfully be accounted for by adding constraints to the observer algorithm. Considering the 1DoF case, the following practical formula can be derived from (2.29) to approximate the lowest observable value of the instantaneous stiffness s based on the approximated value of the system's average linearised stiffness $\mathrm{E}(s)$ and its effective linearised damping ratio, denoted by $\mathcal{D}$:

$$\mathcal{D}^2 \cdot \mathrm{E}(s) < s \,. \tag{2.34}$$

The dynamic behaviour of typical machines and machine components corresponds to averaged damping ratios between 0.01 and 0.1 (Dresig & Fidlin, 2014), (Schlecht, 2009). This means according to (2.34) that in case of typical engineering structures the lowest allowed local effective stiffness s of the nonlinear RFS is about 0.01^2 to 0.1^2 times the average stiffness of the system. This enables that even a system including backlash, which indicates zero stiffness in a significant amplitude range, can be identified with good accuracy using the constrained observer technique (q.v. Section 3.1.2). Furthermore if such minor drawbacks are not affordable, one can still use displacement measurements as an alternative for even better result quality without restrictions on the effective stiffness.

(2) *Is it possible to reduce the number of instrumental variables?*

Although the initial a priori process model (2.3) only provides observability for displacement measurement, it seems to have an advantage in comparison to the final model structure (2.5): It needs only one augmented state f to generate samples of coupled RFSs. This property would indeed reduce the computational costs of the identification algorithm. To discuss this topic let us consider the following three alternative exploratory RFS models with only one augmented state:

$$c(z,v) := sz\,, \tag{2.35}$$

$$c(z,v) := sv\,, \tag{2.36}$$

$$c(z,v) := s(z^2 + \tau^2 v^2)\,, \tag{2.37}$$

where τ is a positive time constant that is necessary to match the units of z and v. In this case s does not represent an effective stiffness any more. It is an instrumental variable that relates the restoring force to the other state variables of the system. The models (2.35) and (2.36) are of the simplest form one can think about. Unfortunately they suffer from a major drawback: As already mentioned in Section 2.1.2, the main challenge of the presented nonparametric approach is that the observer algorithm has to track the fast variation of the augmented states over time caused by the nonlinearities of the observed system. Considering mechanical structures with both elastic and dissipative properties in case of $c(z,v):=sz$ the value of s approaches infinity every time z approaches zero in order to compensate the missing dissipative force. The same holds for $c(z,v):=sv$ as v approaches zero while the elastic force is not equal to zero. This behaviour introduces undesirable artificial oscillations in s, which makes it impossible for the observer to track its value with acceptable accuracy. In fact even the direct estimation of the restoring force, i.e. $c(z,v):=f$, is disadvantageous in this regard, since, while the oscillations of s and d in (2.5) arise solely from nonlinearities, the time history of f is already oscillatory for a vibrating linear system. To avoid this problem one can take the RFS model (2.37), where s never approaches infinity, since the square of the distance from the origin of the phase plane, i.e. z^2+v^2, remains positive as long as the system is in motion. However, the observability investigation of this a priori model structure shows that the system is technically unobservable for $y \epsilon \{v, \dot{v}\}$. A detailed symbolic observability analysis of all three exploratory models can be found in Appendix C, which reveals that actually all of them suffer from major observability issues. There might be some more complex RFS model structures that can solve these issues. However, complex (strongly nonlinear) a priori models cause significantly increasing computational time of observer algorithms, which makes a simple model such as (2.5) highly attractive. This important aspect is discussed in Section 1.6.3 and 2.2.3.

(3) *Is it possible to support the observer convergence by a better model?*

The chosen a priori process model (2.5) assumes constant effective stiffness and effective damping. This assumption is obviously not true, since the aim of the current thesis is to identify nonlinear systems. This means that the convergence of the augmented states to their real values has to be achieved completely via the correction term of the observer equation (1.24), which requires proper tuning of the observer algorithm (q.v. Section 2.2.4). Although this is the typical model choice for the augmented states if no a priori information about their variation over time is available (Lourens, Reynders, De Roeck, Degrande, & Lombaert, 2012), the literature on target tracking offers an alternative approach that can enhance observer convergence. The main idea of this approach is to introduce higher order time derivatives of the existing augmented states as additional instrumental variables (Simon, 2006), (Wu, Lin, Han, & Li, 2010), (Young, 2000). Therefore, it needs to be clarified why the current model does not utilise this option. The idea of introducing time derivatives of the augmented states is based on the expectation that the higher the order of the derivative is, the smoother its time variation gets, and therefore the easier it becomes to track its values. This is a justified expectation for approximately polynomial time variations. However, the current augmented states (**s** and **d**) exhibit oscillatory behaviour in case of vibrating systems including nonlinear RFSs. The introduction of additional time derivatives is obviously not useful for oscillatory variables, since differentiation does not result in smoother characteristics in such cases.

The above reasoning aims to justify the author's choice of the a priori model.

2.2 Observer design for nonparametric estimation

In Section 2.1 an appropriate form of the system equations, i.e. (2.31) and (2.33), has been derived that ensures the observability of the augmented state vector. The current section presents the observer algorithm that is able to synchronise the derived system model to the measurement signals in order to generate RFS samples. This mainly consists of generating a time sequence of the synchronisation gain **K** that satisfies (1.26) using a nonlinear observer (q.v. Section 2.2.3 and 2.2.4), but also includes some further algorithm refinements (q.v. Section 2.2.5 and 2.2.6). Since the synchronisation of a virtual model to noisy real-life measurements essentially changes the way the state space model needs to be treated, some supplementary discussions (q.v. Section 2.2.1 and 2.2.2) are necessary prior to the investigation of the observer algorithm itself. These discussions explain the author's choice of the implemented type of observer, the Extended Kalman Filter, and they introduce the probabilistic point-of-view that plays a crucial role throughout the rest of the thesis.

2.2.1 Probabilistic aspects

Recall from equation (1.24) that due to the correction term in the observer equation the measurement signal **y** is embedded into the modified equation of the virtual system, which leads to two main effects. On the one hand, the process model is recursively corrected by the information about the real system that is carried in **y**. Unfortunately, on the other hand, the once deterministic virtual model is also affected by the undesired part of **y**, which is called the measurement noise vector, denoted by **r**. This undesired signal component is caused by a chain of deterministic processes that arise from the experiment setup and are in most cases unknown and extremely complex. Therefore, it is justified (and also convenient) to consider them as one overall random white noise process that is described by its bias vector and its covariance matrix. Significantly biased measurements can not be handled by the synchronisation algorithm. They can drastically decrease result quality or even cause observer divergence. In most situations **y** can be kept unbiased by proper experiment setup. If it is not the case, then if the bias of the noise is known (or detectable based on a priori expectations on the measurement signals), then it can be compensated before **y** is injected into the observer. Unfortunately this does not hold for the covariance, that is practically speaking the actual noisiness of the signal. A priori noise-filtering the signal can alter the useful higher harmonic components, which one investigating strongly nonlinear systems is especially interested in. Hence, unbiased noise is an inevitable part of every measured signal and has to be accounted for in the observer algorithm. Since today's equipment record digitally sampled signals, the superimposed noise component is considered as a discrete-time process defined by its discrete-time covariance matrix, denoted by **R**. In most cases it is sufficient to assume a so called "white noise", which implies two main properties of the noise process: The spatial distribution of its samples has to be unbiased and normal (Gaussian), denoted as $\mathbf{r} \sim N(0,\mathbf{R})$, whereas the sequence of its samples has to be uncorrelated over time. Noise processes with certain correlation over time, so called "coloured noise", can also be handled by the observer algorithm (Simon, 2006), however, this topic is not considered in the current thesis.

The presence of noise implies a crucial change in the way state variables are handled. The probabilistic variation of the injected measurement is carried over to the virtual system, hence, the augmented state vector itself becomes a probabilistic (random) vector **X**, described by its mean vector **x** and its covariance matrix $\mathbf{P_X}$. One single observed value of **X** at a certain time point *i* is only a sample of its probability distribution, denoted by $\mathcal{X}_i$. The mean and the covariance can only be estimated based on a finite number n_s of these samples using the following two formulas (Meyer, 2003).

$$\mathbf{x} \approx \frac{1}{n_s} \sum_{i=1}^{n_s} \mathcal{X}_i \quad \textbf{(2.38)}$$

$$\mathbf{P_X} \approx \frac{1}{n_s - 1} \sum_{i=1}^{n_s} (\mathcal{X}_i - \mathbf{x})(\mathcal{X}_i - \mathbf{x})^{\mathrm{T}} \quad \textbf{(2.39)}$$

The so called statistical convergence of such a sampling procedure is illustrated in Figure 2.1 for a normally distributed scalar state X with a mean value of 1 and different variance values P_X, denoted as $X\sim\mathrm{N}(1,P_X)$. It can be seen that the *eRS%* error (q.v. Appendix B) of the mean value strongly depends on the true value of the variance.

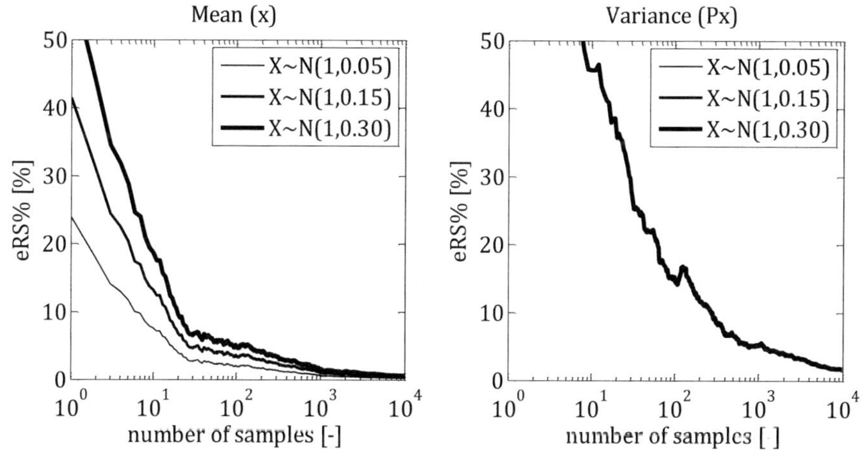

Figure 2.1: Statistical convergence of the mean (left) and variance (right) of the normally distributed scalar variable *X* for different variance values. Convergence is assessed by the *eRS%* deviation (q.v. Appendix B) from the true values averaged over 30 sampling runs.

The error of the variance is independent from its true value, however, its convergence is much slower than that of the mean value. But for all that, the most important message of Figure 2.1 is that regardless of the performance of the observer algorithm itself the expectable result quality (e.g. of the identified RFS) has a natural lower bound that increases as the number of available samples decreases. As a rule of thumb at least 1000 samples are necessary to reach an error level of an estimated mean value at the order of magnitude of 1 %.

2.2.2 Comparison of probabilistic transformations

In order to properly account for noisy measurement signals according to the requirements from Section 1.1, the Kalman Filter (KF) is chosen in the present thesis to synchronise the a priori virtual model (derived in Section 2.1.2) to the real system. As discussed in Section 1.6.2 and 1.6.3, the choice of the implemented probabilistic transformation Φ crucially determines the accuracy of the KF in case of nonlinear models, such as the derived a priori model. The current section presents a short comparison of the three types of Φ from Section 1.6.2 to justify the author's choice of type LinT for nonparametric identification in Section 2.2.3. To do so, let us consider a particular random vector **X** with the following properties:

$$\mathbf{X} = \begin{bmatrix} X_1 \\ X_2 \\ X_3 \end{bmatrix} \sim \mathrm{N}\left(\mathbf{x} = \begin{bmatrix} 1 \\ 2 \\ 2 \end{bmatrix}, \mathbf{P}_\mathbf{X} = \begin{bmatrix} 0.15 & 0.15 & 0.15 \\ 0.15 & 0.3 & 0.2 \\ 0.15 & 0.2 & 0.3 \end{bmatrix} \right) \tag{2.40}$$

The performance of the three algorithms is compared in Figure 2.2 and Figure 2.3 for the bilinear $Y=X_1X_2$ and trilinear $Y=X_1X_2X_3$ nonlinear transformations respectively based on the mean y and the variance P_Y of the transformed random variable Y. The MCT curves represent the averaged results of 30 independent Monte Carlo sampling runs.

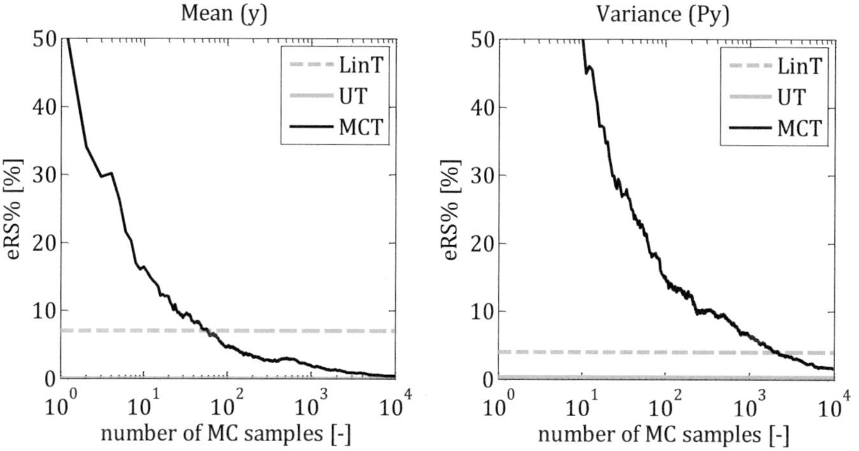

Figure 2.2: Estimation error of the mean (left) and variance (right) of the random variable Y, given by the transformation $Y=X_1X_2$. A MCT with $3 \cdot 10^5$ samples is taken as the reference for the *eRS%* deviation (q.v. Appendix B).

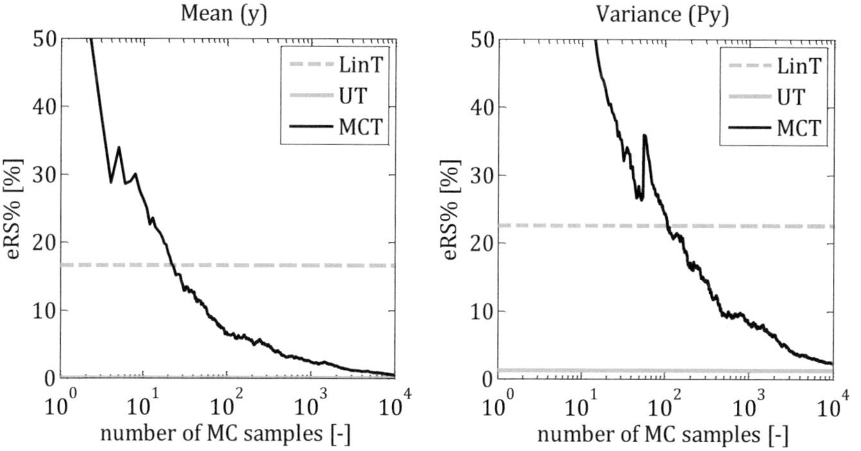

Figure 2.3: Estimation error of the mean (left) and variance (right) of the random variable Y, given by the transformation $Y=X_1X_2X_3$. A MCT with $3 \cdot 10^5$ samples is taken as the reference for the *eRS%* deviation (q.v. Appendix B).

Notice that the mean value of Y determined by LinT deviates from its true value. Since the calculation of the mean value is completely deterministic in case of LinT, q.v. Equation (1.35), this shows that simply ignoring the noisiness of measurement signals can cause systematic error of the identified system properties in case of nonlinear models. Regarding the comparison of the different algorithms it can be seen that for the particular nonlinearities the MCT requires approximately 10^4 samples to reach the performance of the UT, whereas the UT requires less than 10 sigma points due to the low dimension of **X**. Hence, it can be concluded that for problems of small and middle dimension (dim(**X**)<100) with moderate nonlinearities (and approximately Gaussian distributions) the UT is superior over the MCT. Furthermore, it can be observed that, though the UT always provides better results than the LinT, in case of the bilinear function the LinT shows good performance as well. This is a very important conclusion, since the a priori system model introduced in Section 2.1.2 includes solely bilinear nonlinear terms.

2.2.3 Implementation of the Kalman Filter

In this section the implementation of the KF (q.v. Section 1.6.3) for the general system model, proposed in Section 2.1.2, is presented. The procedure estimates the probabilistic augmented state vector $\mathbf{X}_a$ at every measurement time step with the following structure:

$$\mathbf{X}_a = \begin{bmatrix} \mathbf{Z} \\ \mathbf{V} \\ \mathbf{S} \\ \mathbf{D} \end{bmatrix} \sim \left(\mathbf{x}_a = \begin{bmatrix} \mathbf{z} \\ \mathbf{v} \\ \mathbf{s} \\ \mathbf{d} \end{bmatrix}, \mathbf{P}_{\mathbf{X}a} = \begin{bmatrix} \mathbf{P_Z} & \mathbf{P_{ZV}} & \mathbf{P_{ZS}} & \mathbf{P_{ZD}} \\ \cdot & \mathbf{P_V} & \mathbf{P_{VS}} & \mathbf{P_{VD}} \\ \cdot & \cdot & \mathbf{P_S} & \mathbf{P_{SD}} \\ \cdot & \cdot & \cdot & \mathbf{P_D} \end{bmatrix} \right), \tag{2.41}$$

where $\mathbf{x}_a$ and $\mathbf{P}_{\mathbf{X}a}$ denotes the mean vector and the covariance matrix of $\mathbf{X}_a$ respectively. The recursive algorithm is based on the general KF formulation given by the equations (1.45), (1.46), (1.47) and (1.48). Therefore, arbitrary types of the probabilistic transformation Φ can be deployed. Of course the choice of Φ has to be made carefully according to the complexity of the a priori system model. In the present study the EKF (i.e. type LinT) and the SR-UKF (i.e. type UT) have been tested. It is shown in further chapters, that the proposed approach is designed to enable the use of the EKF regardless of the nonlinearities involved in the RFSs that are to be identified, as long as the measurement equation (2.33) is linear. This is explained by Section 2.2.2, where it is shown that the bilinear nonlinearity of the process model (2.31) can be described with good accuracy using Φ of type LinT. Deploying the EKF results in a computationally efficient algorithm, which is in accordance with the requirements from Section 1.1. The SR-UKF is only used as a reference for the EKF in order to prove its sufficiency. Nevertheless, the SR-UKF should still be kept in mind as an alternative for the cases, when:

- the measurement function (**h**) is significantly nonlinear;
- the a priori model (2.31) has to be extended with additional strongly nonlinear terms;
- the system model is given in form of a black-box code (generated, e.g., by some software for dynamic simulation), which does not allow the symbolic calculation of the Jacobians that are needed for the EKF;
- the model development is in exploratory phase, where the ease of model changes is more important than computational time.

Technically the complete recursive algorithm of the EKF is presented in Section 1.6.2 and 1.6.3. However, one step of the recursion, namely the time update (1.45), needs some extra discussion. Carrying out the probabilistic transformation of the augmented state vector through the integral of the process equation (2.31) has to be treated carefully. The first non-trivial task is the transformation of the mean vector (1.35), i.e. the time-integration itself. This integral in general can not be given in an explicit form and therefore has to be approximated using a proper integration scheme. Though the chosen general model is linear between measurement time steps (since **s** and **d** are only changed in the correction step), in order to keep the algorithm flexible and consistent with the simulation of the numerically generated measurements of nonlinear systems (q.v. Chapter 3 and 4), it is reasonable to apply an integration strategy that is designed for nonlinear systems. A huge number of such integration algorithms are available, and the optimal choice depends crucially on the structure of the system model (Walter, 2014). As long as the process model is given by relatively simple ODEs, which holds for (2.31), a simple explicit Runge-Kutta (RK) algorithm is sufficient in most cases. Normally the higher the order of the RK, the larger the affordable integration step size becomes, and one is able to find an optimal order that results in the shortest possible integration time. This principle can not be applied for the KF, since the maximum allowed integration time step size is given by the sampling frequency of the measurement signals. If an experiment is carried out with the purpose of system identification, a fine resolution of the oscillations is advantageous, which typically leads to a measurement sampling rate that is about ten to twenty times the highest expected frequency in the signal. Due to this, in most cases a relatively low order RK is sufficient to achieve good integration quality. Additionally it should be kept in mind that, since we attempt to observe nonlinear systems, the true value of **s** and **d** will change over time. This means that the a priori model diverges from reality during the integration time step, which reduces the value of a precise integration algorithm. Therefore, throughout the current study a third-order explicit Runge-Kutta integration scheme (RK3) is used, which leads to the following formula for the time update of the state vector's mean in the EKF algorithm:

$$\hat{\mathbf{x}}_{a,i}^{-} = \hat{\mathbf{x}}_{a,i-1}^{+} + \int_{t_{i-1}}^{t_i} \mathbf{a}(\hat{\mathbf{x}}_a, \mathbf{u})dt \approx \hat{\mathbf{x}}_{a,i-1}^{+} + \frac{\Delta t(\mathbf{a}_I + 4\mathbf{a}_{II} + \mathbf{a}_{III})}{6}, \tag{2.42}$$

where Δt denotes the measurement sampling time and the terms $\mathbf{a}_j$ are given as

$$\mathbf{a}_I := \mathbf{a}\left(\hat{\mathbf{x}}_{a,i-1}^{+}, \mathbf{u}_{i-1}\right), \tag{2.43}$$

$$\mathbf{a}_{II} := \mathbf{a}\left(\hat{\mathbf{x}}_{a,i-1}^{+} + (\Delta t \mathbf{a}_I)/2, (\mathbf{u}_{i-1} + \mathbf{u}_i)/2\right), \tag{2.44}$$

$$\mathbf{a}_{III} := \mathbf{a}\left(\hat{\mathbf{x}}_{a,i-1}^{+} - \Delta t \mathbf{a}_I + 2\Delta t \mathbf{a}_{II}, \mathbf{u}_i\right). \tag{2.45}$$

The second nontrivial task during the time update step of the EKF is the approximation of the transformed covariance matrix of the augmented state, which requires the Jacobian of the time integral of the process equation that (similar to the integral itself) in general can not be given exactly and therefore has to be approximated. To do so, an explicit algorithm based on the Jacobian of the process function, denoted by **A**, is deployed. According to the current a priori process model (2.31), **A** is of the form (2.46). It is apparently not a constant matrix. Therefore, a trapezoidal formula of the form (2.47) has been introduced in the present study, which allows a good approximation of the Jacobian of the process function's time integral. This Jacobian is then used for the transformation of the state covariance matrix during the time update in the EKF according to the linearised formula (1.36).

$$\mathbf{A}_i = \begin{bmatrix} \mathbf{0} & \mathbf{I} & \mathbf{0} & \mathbf{0} \\ -\mathbf{M}^{-1}\mathrm{diag}(\mathbf{s}_i) & -\mathbf{M}^{-1}\mathrm{diag}(\mathbf{d}_i) & -\mathbf{M}^{-1}\mathrm{diag}(\mathbf{z}_i) & -\mathbf{M}^{-1}\mathrm{diag}(\mathbf{v}_i) \\ \mathbf{0} & \mathbf{0} & \mathbf{0} & \mathbf{0} \\ \mathbf{0} & \mathbf{0} & \mathbf{0} & \mathbf{0} \end{bmatrix} \tag{2.46}$$

$$\left.\frac{\partial\left(\hat{\mathbf{x}}_{a,i-1}^{+} + \int_{t_{i-1}}^{t_i} \mathbf{a}(\hat{\mathbf{x}}_a, \mathbf{u})dt\right)}{\partial \mathbf{x}_a}\right|_{i-1} \approx \mathbf{I} + \frac{\Delta t\left(\mathbf{A}(\hat{\mathbf{x}}_{a,i-1}^{+}) + \mathbf{A}(\hat{\mathbf{x}}_{a,i}^{-})\right)}{2} \tag{2.47}$$

If the general process equation (2.31) has to be extended with strong nonlinearities or additional algebraic equations (i.e. if one has to deviate from the problem formulation from Section 1.3) then the integration over time might require more sophisticated algorithms. These for e.g. can include event-handling (Stamm, 2011) or can accommodate DAEs (Fischer, 2013). In such cases it is convenient to use the SR-UKF (instead of the EKF) in combination with a black-box system model that includes the appropriate integrator as well. Such models can be generated by most of the state of the art dynamic simulation software.

2.2.4 Tuning for nonparametric identification

So far the adaption of the general EKF equations to the particular a priori model has been presented (q.v. Section 2.2.3). As mentioned in Section 1.6.3, in order to run the EKF the initial a posteriori distribution of $\mathbf{X}_a$ has to be characterised via $\hat{\mathbf{x}}_{a,0}^{+}$ and $\widehat{\mathbf{P}}_{\mathbf{X}a,0}^{+}$, and the process and measurement discrete-time covariance matrices, i.e. $\mathbf{Q}$ and $\mathbf{R}$, have to be defined as well. Let us start with the latter, which is a determinable property of the measurement signal. Ideally $\mathbf{R}$ should be available from specific noise analysis tests that have been carried out on the particular test rig. Since it is not always the case, the identification procedure has to be prepared for the determination of $\mathbf{R}$ from the same measurements that are used for the identification itself, i.e. $\mathbf{y}$. To do so, in the current study the three-point Central Moving Average (q.v. Appendix E) of $\mathbf{y}$, denoted by $\mathbf{y}^{\mathrm{CMA3}}$, has been subtracted from its original values to generate an approximation of the noise sequence. This is then statistically analysed using the formula

$$\mathbf{R} \approx \frac{3}{2n_s} \sum_{i=1}^{n_s} \left(\mathbf{y}_i - \mathbf{y}_i^{\mathrm{CMA3}}\right)\left(\mathbf{y}_i - \mathbf{y}_i^{\mathrm{CMA3}}\right)^{\mathrm{T}}. \qquad \textbf{(2.48)}$$

Although this is a quick and dirty solution, it proved to deliver sufficiently accurate estimates. The initial guess of the state vector, i.e. $\hat{\mathbf{x}}_{a,0}^{+}$, can be used to involve any a priori knowledge about the states, therefore, it is defined as (2.49), where E(...) denotes the expectation operator. Especially the expected average stiffness and damping values, i.e. $\mathrm{E}(\mathbf{S})$ and $\mathrm{E}(\mathbf{D})$, can be advantageous to increase the initial convergence of the synchronisation. In case of oscillatory system behaviour, where the time average of displacements and velocities is approximately zero, it is reasonable to define $\mathrm{E}(\mathbf{Z}_0)$ and $\mathrm{E}(\mathbf{V}_0)$ as zero vectors. Since the initial correlation between the state variables is typically unknown, $\widehat{\mathbf{P}}_{\mathbf{X}a,0}^{+}$ can be defined as a diagonal matrix with variance values representing the range in which the specific initial state values are expected to be located. If the initial state is completely unknown, assuming a standard deviation of 100% of the expected initial mean values proved to be a good choice. This leads to an initial state covariance matrix of the form (2.50), where the square operation denotes taking the square of each vector coordinate element-wise. Notice that even for symmetrical oscillatory behaviour $E(\mathbf{Z}^2)$ and $E(\mathbf{V}^2)$ will not be zero. This is important first, because the state covariance matrix is per definition always positive definite (and symmetric), and second, because zero variance would mean completely certain initial values of the corresponding states.

$$\hat{\mathbf{x}}_{a,0}^{+} := [\, E(\mathbf{Z}_0)^{T} \quad E(\mathbf{V}_0)^{T} \quad E(\mathbf{S})^{T} \quad E(\mathbf{D})^{T} \,]^{T} \tag{2.49}$$

$$\widehat{\mathbf{P}}_{\mathbf{x}a,0}^{+} := \mathrm{diag}([\, E(\mathbf{Z}^2)^{T} \quad E(\mathbf{V}^2)^{T} \quad E(\mathbf{S}^2)^{T} \quad E(\mathbf{D}^2)^{T} \,]) \tag{2.50}$$

Finally the process covariance matrix $\mathbf{Q}$ has to be defined. It remains as the only real tuning factor of the KF that has a crucial influence on the convergence of the synchronisation. It is a symmetric matrix of size $n_a \times n_a$ that describes the uncertainty of the corresponding n_a equations of the process model with respect to their symbolic structure and constant parameters. The classical purpose of an observer (and therefore of the KF as well) is to compensate the error between $\hat{\mathbf{x}}_{a,0}^{+}$ and the a priori unknown real initial state $\mathbf{x}_{a,0}$. It implies a system model that is at least expected to be correct (regarding its structure and parameters). Under such conditions $\mathbf{Q}$ can theoretically be set to zero, and the virtual model's behaviour, represented by the estimated measurement signal, will converge to the real system's behaviour. Small non-zero diagonal values in $\mathbf{Q}$ can be used to "keep the KF alive", i.e. to force the algorithm to keep on compensate deviations that are caused by unexpected modelling errors and measurement disturbances. These classical conditions are not satisfied in the current case. The instrumental variables, i.e. $\mathbf{s}$ and $\mathbf{d}$, are expected to vary very quickly over time, which is however not described by the process model due to the a priori unknown dynamics. Therefore, an unusual tuning strategy has to be introduced in the present thesis, where the diagonal values of $\mathbf{Q}$ related to the augmented part of the state vector are set to high values to indicate the uncertainty of the corresponding rows of the process model. These diagonal values are tuned by the exponents q_s and q_d that allow the separate convergence adjustment of the instantaneous effective stiffness and damping respectively. In order to make the increase rate of the process uncertainty independent of the measurement sampling frequency, $\mathbf{Q}$ is set proportional to the sampling time step Δt. This leads to the following formula for the discrete time process covariance matrix:

$$\mathbf{Q} := \Delta t \cdot \mathrm{diag}([\,\mathbf{0}_{1\times N} \quad \mathbf{0}_{1\times N} \quad 10^{q_s} \cdot \mathrm{E}(\mathbf{S}^2)^{\mathrm{T}} \quad 10^{q_d} \cdot \mathrm{E}(\mathbf{D}^2)^{\mathrm{T}}\,])\,. \qquad \textbf{(2.51)}$$

In order to carry out the tuning, a filter convergence plot is proposed in the current study as a helpful tool for finding the proper values of q_s and q_d. To evaluate the convergence of the KF, the synchronisation residual is calculated over time in a normalised mean square (*eMS*) form. Due to the predictor-corrector formulation of the KF, two different residuals with different results can be evaluated. The "prediction error", denoted by e_p, is calculated using the a priori estimate of the measurement $\hat{\mathbf{y}}$, whereas the "synchronisation error", denoted by e_s, is determined by the a posteriori measurement estimate $\hat{\mathbf{y}}^+$. As a reference, these errors are compared to the *eMS* form of $\mathbf{R}$, called "measurement error", denoted by e_m. The precise definition of these error quantities according to Appendix B is given as

$$e_m := eMS^{0,i_{max}}_{(\mathbf{y}-\mathbf{y}^{\mathrm{CMA3}}),\mathbf{y}}\,, \qquad e_p := eMS^{i-n_s,i}_{\hat{\mathbf{y}},\mathbf{y}}\,, \qquad e_s := eMS^{i-n_s,i}_{\hat{\mathbf{y}}^+,\mathbf{y}}\,. \qquad \textbf{(2.52)}$$

The measurement error is calculated a priori using the complete measurement signal, whereas e_p and e_s are calculated during the KF run as moving errors with a sample window size of n_s, which should be set wide enough to average over all dominant oscillation frequencies of the measurement signal. The calculation of the vector $\hat{\mathbf{y}}^+$ is not a standard component of the KF algorithm. It is carried out via an extra call of the measurement equation after the correction step (1.48), which has been implemented in the present study in order to enable some useful analysis features that are presented in the following.

If the system model is non-augmented or the state augmentation technique is used for the identification of constant or slowly varying parameters, the typical tuning of the KF is sufficient. It results in the synchronisation error converging to the measurement error, which provides a smooth noisefree estimate of the augmented states (q.v. Figure 2.4). The prediction error also converges to the measurement error, since the model is corrected over successive iterations and hence predicts accurate noisefree measurement values. This practice can be observed in (Wu & Smyth, 2007), (Chatzi & Smyth, 2009) and (Kolansky & Sandu, 2012), where constant parameters of mechanical systems are estimated using different types of nonlinear KFs.

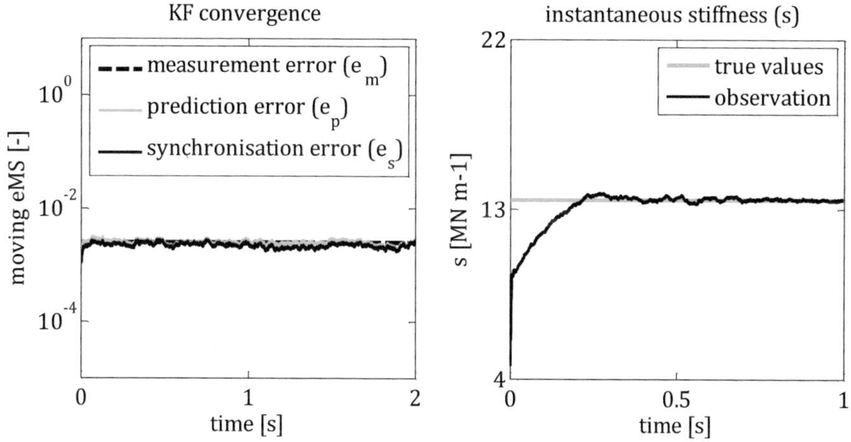

Figure 2.4: Classical tuning of the KF for the estimation of constant augmented states on the example of a virtual frequency sweep measurement of a 1DoF oscillator with constant stiffness and damping. q_s and q_d have been set to -2.

To demonstrate the necessity of an alternative tuning strategy for the current nonparametric problem formulation let us take the numerical example of the 1DoF nonlinear oscillator from Section 3.1.1. Figure 2.5 depicts the result of applying the classical approach, i.e. tuning the synchronisation error to the level of the measurement error. Due to a priori unknown time-variation of the augmented states, the model is invalid at every time step of the KF run, and therefore the prediction error remains higher than the measurement error. This strategy works as a low-pass filter acting on the estimated augmented states. In the present case this not only filters the measurement noise but also useful high-frequency information about the instantaneous stiffness and damping (q.v. Figure 2.5 right). Therefore, in the current study an optimised tuning is proposed to handle the oscillations of the augmented states. The KF is tuned to the highest stable synchronisation gain possible, which results in the synchronisation error reaching a much lower level than the measurement error (q.v. Figure 2.6 left). This leads on the one hand to the synchronisation of the model to the measurement noise, on the other hand to a significant reduction of the instrumental variables' bias (q.v. Figure 2.6 right).

In praxis, increasing q_s and q_d over a limit leads to an unstable observer as a result of the following two effects. First, it leads to a bad conditioning of $\mathbf{P}_{\mathbf{x}a}$, which can cause numerical problems. This can be improved using square-root forms of the KF (such as the SR-UKF) or by scaling the augmented states by their expected mean values and estimating their normalised coefficients. The latter has been implemented in the current study (q.v. Chapter 3). Second, extremely high values of $\mathbf{Q}$ lead to a complete neglect of the process model, which causes the numerical loss of observability. This forms a physical limit of the convergence speed. Figure 2.7 shows how the bias and the variance of s(t) changes as the values of q_s and q_d are increased. It is interesting that there is not only an upper bound of the convergent region, but there is also an additional divergent region in the middle range of the covariance exponents. This phenomenon however has not been deeper investigated in the present study, since it has no influence on the preferred tuning strategy.

During the current investigations (q.v. Chapter 3 and 5) finding the highest possible q_s and q_d was carried out by setting their values to a level, where the synchronisation error becomes lower than the measurement error. This was followed by successively increasing the values until the threshold of instability is reached. For systems with oscillations of interest in the range of 50 Hz to 500 Hz the chosen values were between 2 and 4.

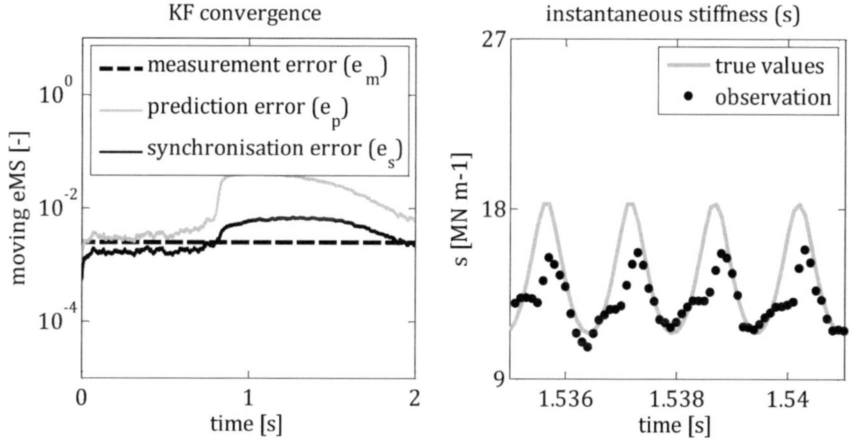

Figure 2.5: Classical tuning of the KF in case of quickly varying augmented states on the example of the virtual 1DoF nonlinear oscillator from Section 3.1.1. q_s and q_d have been set to 1.

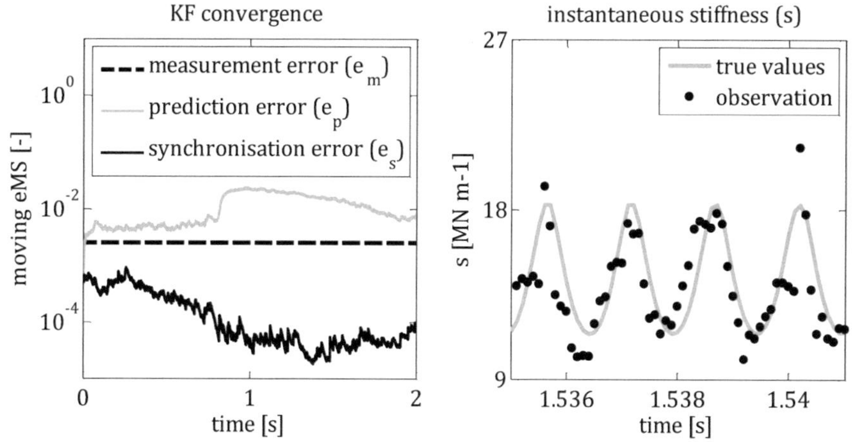

Figure 2.6: Optimised tuning of the KF for quickly varying augmented states on the example of the virtual 1DoF nonlinear oscillator from Section 3.1.1. q_s and q_d have been set to 3.

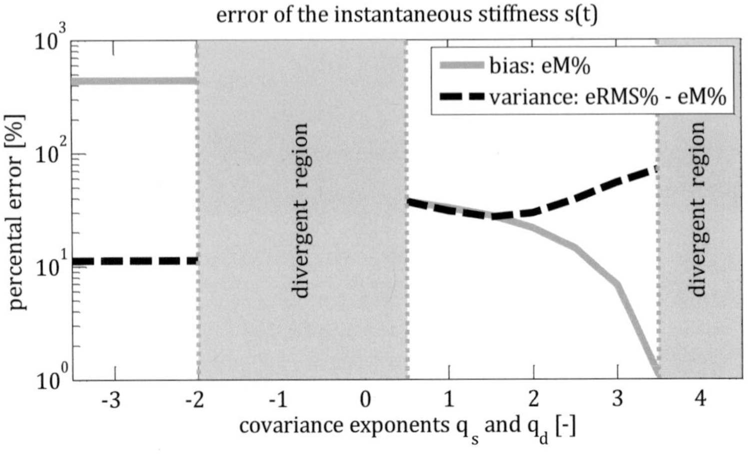

Figure 2.7: Bias and variance of $s(t)$ (based on *eM%* and *eRMS%* according to Appendix B) on the example of the virtual 1DoF nonlinear oscillator from Section 3.1.1 using the EKF. q_s and q_d have been increased simultaneously.

Although the introduced tuning strategy enables the observation of fast oscillations of the instrumental variables (**s** and **d**), it also causes the estimates to be significantly corrupted by noise (q.v. Figure 2.6 and Figure 2.7). This makes an a posteriori statistical analysis of the observed data necessary, which is proposed in Section 2.3.

2.2.5 Constraints

It has been shown in Section 2.1.2 that in case of velocity and acceleration measurements the system loses its observability if the estimated values of the effective stiffness cross a certain lower bound, which is given by (2.29) for a 1DoF system. In order to guarantee the convergence of the KF even in case of small instantaneous stiffness values, additional inequality constraints have to be implemented in the algorithm. The current general a priori process model (2.31), which has been defined according to the problem formulation from Section 1.3, is given by a set of unconstrained ODEs. Once this model is extended by additional state constraints, the formulation of the KF from Section 2.2.3

needs to be reviewed. It has been mentioned in Section 2.2.3 that the most general and convenient way of applying a KF for constrained systems is the direct implementation of the constraints in the process equation in combination with the SR-UKF algorithm. There are two reasons, why this solution is disadvantageous in the current case. First, according to the requirements from Section 1.1, this study aims to establish a computationally fast algorithm, therefore, the EKF should remain deployable. Second, the necessity of constrained instrumental variables does not arise from the physical model directly. It is related to the chosen identification algorithm, and therefore should be treated separately without the modification of the process model. Fortunately there are some techniques that allow the incorporation of simple equality and inequality constraints in the KF (Simon, 2006). One of these techniques, the so called Density Function Truncation, is especially remarkable due to its probabilistic manner of treating inequality constraints. In this method the probability distribution of the concerned states is truncated between the lower and the upper bounds instead of deterministically applying the constraints to the mean value of the state variables. This is much more appropriate in case of stochastic estimation, since the state covariance matrix is adjusted properly as well. An exact formula of this approach for Gaussian distributions is presented in (Simon & Simon, 2006). The constraining strategy that has been implemented in the present thesis follows this idea. It is however formulated by means of the Unscented Transformation (q.v. Section 1.6.2) in order to keep the algorithm general and more flexible.

Considering the EKF, the only step of the recursive algorithm, where the instrumental variables ($\mathbf{s}$ and $\mathbf{d}$) can change their values, is the correction step (1.48), since the a priori system model assumes them to be constant. Therefore, it is reasonable to include the constraint step in the recursion between the correction step (1.48) and the subsequential prediction step (1.45). The uncoupled inequality constraint of the a posteriori state estimate $\hat{\mathbf{x}}^+$ with lower bound $\mathbf{x}^{\mathrm{L}}$ and upper bound $\mathbf{x}^{\mathrm{U}}$ is given by the vector function $\boldsymbol{\kappa}$ as

$$\hat{\mathbf{x}}^* = \boldsymbol{\kappa}(\hat{\mathbf{x}}^+, \mathbf{x}^{\mathrm{L}}, \mathbf{x}^{\mathrm{U}}) := \left[\begin{cases} x_1^{\mathrm{U}} & , \hat{x}_1^+ \geq x_1^{\mathrm{U}} \\ \hat{x}_1^+ & , x_1^{\mathrm{L}} < \hat{x}_1^+ < x_1^{\mathrm{U}} \\ x_1^{\mathrm{L}} & , \hat{x}_1^+ \leq x_1^{\mathrm{L}} \end{cases}, \ldots, \begin{cases} x_{n_a}^{\mathrm{U}} & , \hat{x}_{n_a}^+ \geq x_{n_a}^{\mathrm{U}} \\ \hat{x}_{n_a}^+ & , x_{n_a}^{\mathrm{L}} < \hat{x}_{n_a}^+ < x_{n_a}^{\mathrm{U}} \\ x_{n_a}^{\mathrm{L}} & , \hat{x}_{n_a}^+ \leq x_{n_a}^{\mathrm{L}} \end{cases} \right]^{\mathrm{T}}, \tag{2.53}$$

where $\hat{\mathbf{x}}^*$ denotes the constrained state estimate. Applying this constraint in a probabilistic manner can be formulated as the probabilistic transformation Φ of the mean and variance of the a posteriori augmented state ($\hat{\mathbf{x}}_a^+$ and $\widehat{\mathbf{P}}_{\mathbf{X}a}^+$) through the function $\boldsymbol{\kappa}$. Since this function can not be linearised properly, the UT approach is deployed for the transformation, which is then given as

$$\begin{bmatrix} \hat{\mathbf{x}}_a^* & \widehat{\mathbf{P}}_{\mathbf{X}a}^* & \widehat{\mathbf{P}}_{\mathbf{X}_a^+\mathbf{X}_a^*} \end{bmatrix} = \Phi_{\boldsymbol{\kappa}}^{\mathrm{UT}}\left(\hat{\mathbf{x}}_a^+, \widehat{\mathbf{P}}_{\mathbf{X}a}^+\right), \tag{2.54}$$

where $\hat{\mathbf{x}}_a^*$ and $\widehat{\mathbf{P}}_{\mathbf{X}a}^*$ denote the mean and variance of the constrained augmented state respectively. These constrained values replace the a posteriori ones in the prediction step (1.45) in the modified KF algorithm. At those points of the following chapters, where the SR-UKF is used as a reference for the EKF, the constraint equation (2.53) is directly implemented in the process model. This means that $\boldsymbol{\kappa}$ is applied in a deterministic way to each a priori sigma point of the state vector, i.e. $\boldsymbol{\mathcal{X}}_j^-$, during the time update step (1.45).

Throughout the current thesis (if not explicitly stated otherwise) the lower and upper bounds are defined as

$$\mathbf{x}^{\mathrm{L}} = \begin{bmatrix} -\boldsymbol{\infty}_{1\times N} & -\boldsymbol{\infty}_{1\times N} & \mathbf{s}_{1\times N}^{\mathrm{L}} & -\boldsymbol{\infty}_{1\times N} \end{bmatrix}^{\mathrm{T}}, \qquad \mathbf{x}^{\mathrm{U}} = \boldsymbol{\infty}_{n_a\times 1}, \tag{2.55}$$

where $-\boldsymbol{\infty}$ (and $\boldsymbol{\infty}$) denote vectors with entries low (and high) enough for the corresponding states to remain unconstrained. $\mathbf{s}^{\mathrm{L}}$ denotes the lower bound vector of the instantaneous stiffness vector $\mathbf{s}$ that is necessary in order to maintain observability in case of velocity and acceleration measurements. Its values can be determined according to (2.34).

2.2.6 Convergence monitoring

The convergence plot has been proposed in Section 2.2.4 as a helpful tool for finding the proper values of q_s and q_d, i.e. for the tuning of the KF. It is however not its only possible application. It is used for two further important purposes in the present study, which are briefly discussed in this section.

In case of complex systems, long measurement sequences or large parameter studies live monitoring of the KF during the estimation is essential. This possibility enables the reduction of unnecessary computation time by terminating the process as soon as convergence issues occur. The convergence plot is perfectly suitable for this purpose, since the prediction error e_p and the synchronisation error e_s can be processed and plotted recursively during the KF run. Let us take the virtual frequency sweep measurement of the 1DoF oscillator from Section 3.1.3 to demonstrate this capability. This system makes it is easy to induce typical convergence problems, due to its strong nonlinearities involving asymmetric stiffness and Coulomb friction. One possible reason of KF divergence is the wrong choice of the covariance exponents (q_s and q_d), which has already been mentioned in Section 2.2.4. Such a situation is depicted in Figure 2.8, where setting the exponents to q_s=q_d=5 results in a divergent KF, which can be detected easily by taking a look at the convergence plot.

Even if the covariance exponents are set to proper values (in this case q_s=q_d=3), such a strongly nonlinear system can lead to further undesired behaviour. This is the case e.g. if the lower bound constraint $\mathbf{s}^L$ of the instantaneous stiffness is deactivated. Such a situation is depicted in Figure 2.9. It can be seen that, though the synchronisation is convergent, the KF becomes inconsistent due to the loss of observability. The stiffness converges to incorrect values around zero. Again the malfunction is indicated clearly on the convergence plot by the unrealistically low level of the synchronisation error e_s.

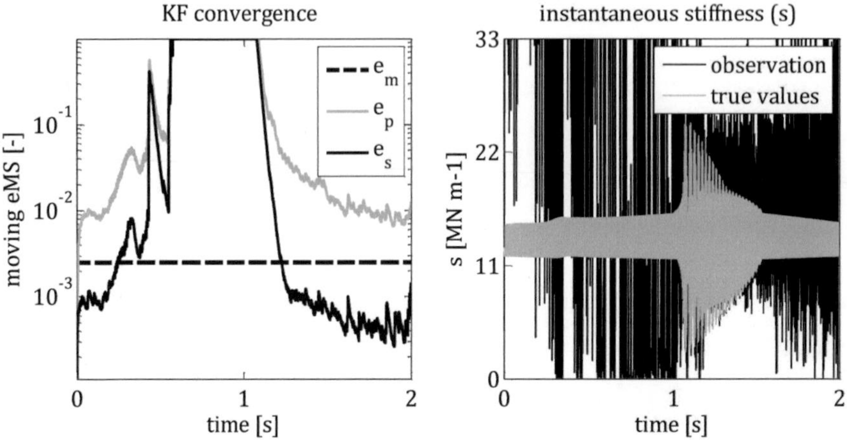

Figure 2.8: Convergence plot (left) and the time sequence of the instantaneous stiffness (right) in case of KF divergence due to improper tuning. Virtual frequency sweep measurement of the strongly nonlinear 1DoF oscillator from Section 3.1.3.

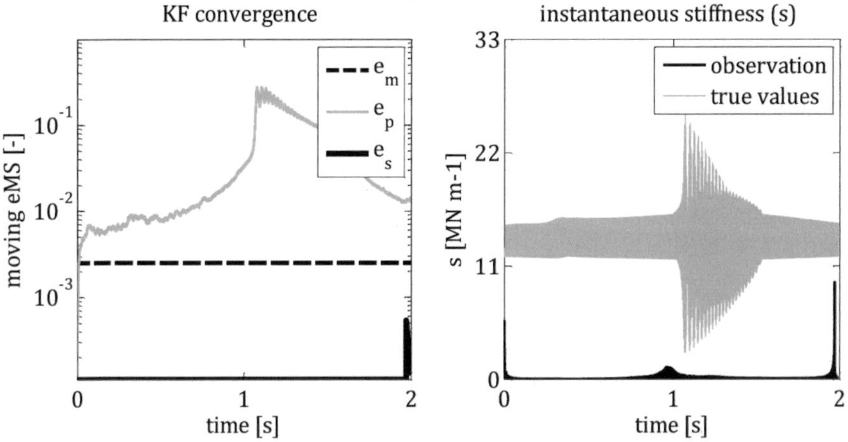

Figure 2.9: Convergence plot (left) and the time sequence of the instantaneous stiffness (right) in case of the inconsistence of the KF caused by the deactivation of the lower bound constraint of the stiffness. Virtual frequency sweep measurement of the strongly nonlinear 1DoF oscillator from Section 3.1.3.

The other important purpose of the convergence plot is its essential role in the a posteriori analysis of the estimated time sequences of the augmented states.

In case of the particular nonlinear system the KF exhibits several short-time convergence issues, even if it is properly tuned, which is depicted in Figure 2.10. The estimated stiffness values show high deviations from the true values in those time ranges, where the synchronisation error e_s reaches high levels. Major deviations occur especially in those regions, where e_s becomes larger than the measurement error e_m. The convergence plot can be used to discard the low quality parts of the estimation results from the postprocessing by setting a properly chosen threshold for e_s. Discarding all state estimates that correspond to $e_s>0.5e_m$ proved to be sufficient for the identification examples that are presented in the following chapters. In this particular example the defined threshold results in the exclusion of approximately the first 0.5 s of the estimated time sequence. Notice that in the time ranges $t\in[0.5,1]$ and $t\in[1.5,2]$ the observed stiffness still deviates significantly from the true values. This error is however nearly unbiased and can be accounted for a posteriori using a proper probabilistic fitting approach, which is proposed in Section 2.3.

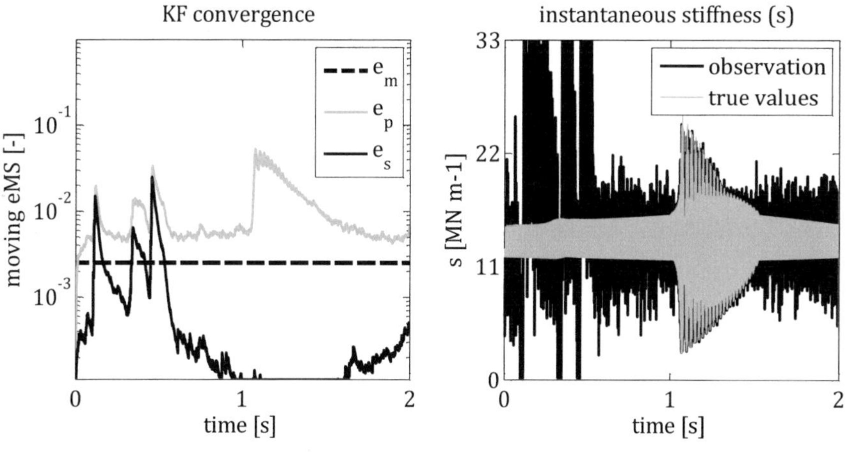

Figure 2.10: Convergence plot (left) and the time sequence of the instantaneous stiffness (right) according to the virtual frequency sweep measurement of the strongly nonlinear 1DoF oscillator from Section 3.1.3.

2.3 Reconstruction of noisefree RFSs

In Section 2.1 an *N*DoF a priori system model (given by (2.31), (2.32) and (2.33)) has been proposed that enables the indirect observation of RFS samples via instrumental variables (**s** and **d**) in order to carry out the nonparametric identification according to the problem formulation from Section 1.3. The implementation of this model in a nonlinear KF algorithm has been presented in Section 2.2, which generates an estimated time sequence of the augmented state's mean vector $\mathbf{x}_a$ and covariance matrix $\mathbf{P}_{\mathbf{x}a}$ in the form (2.41) based on the excitation signal **u** and the measured global behaviour of the real system given by **y**. The elements of the estimated time series can now be assigned to the *N* RFSs, which results in the *j*th local mean vector $\mathbf{x}_{C,j}$ and covariance matrix $\mathbf{P}_{C,j}$ of the form

$$\mathbf{X}_{C,j} := \begin{bmatrix} Z_j \\ V_j \\ S_j \\ D_j \end{bmatrix} \sim \left(\mathbf{x}_{C,j} := \begin{bmatrix} z_j \\ v_j \\ s_j \\ d_j \end{bmatrix}, \mathbf{P}_{C,j} := \begin{bmatrix} P_{Z,j,j} & P_{ZV,j,j} & P_{ZS,j,j} & P_{ZD,j,j} \\ \cdot & P_{V,j,j} & P_{VS,j,j} & P_{VD,j,j} \\ \cdot & \cdot & P_{S,j,j} & P_{SD,j,j} \\ \cdot & \cdot & \cdot & P_{D,j,j} \end{bmatrix} \right), \quad \textbf{(2.56)}$$

where the mean and variance elements correspond to the *j*th and *j,j*th element of the mean vector and covariance matrix in (2.41). Notice that, though the "hat" symbol has been dropped, all quantities in (2.56) represent estimated values. Throughout Section 2.3 this simplified notation is used. Due the assignment of the KF results to local RFSs, the rest of the identification algorithm can be split into *N* independent RFS fitting problems. Therefore, for the rest of the current section the indices "*j*" and "*j,j*" are dropped for the ease of notation. In general the deformation z and rate of deformation v of the RFS do not necessarily equal to particular coordinates of the chosen state-space of the system, but can be given by arbitrary functions of the state vector as $z=\beta(\mathbf{x})$ and $v=\gamma(\mathbf{x})$ (q.v. (1.6) and (1.7)). In such cases (that deviate from the current problem formulation from Section 1.3) z, v and their variances can be generated using a properly chosen type of probabilistic transformation of $\mathbf{x}$ (q.v. Section 1.6.2) through the functions β and γ.

In Subsection 2.3.1 the reconstruction of probabilistic RFS samples based on the local augmented state vector is presented. This is followed by the derivation of an optimal curve fitting algorithm in Subsection 2.3.2 to generate noisefree RFCs to describe additive RFSs. Finally the optimal surface fitting of the samples is discussed in Subsection 2.3.3 that allows the reconstruction of noisefree coupled RFSs.

2.3.1 Reconstruction of RFS samples

Recall from Section 1.4.2 that the ith sample of eRFCs, dRFCs and RFSs are given by the coordinate sets $\{z,f_E\}_i$, $\{v,f_D\}_i$ and $\{z,v,f\}_i$ respectively, to which the following notation is introduced:

$$\boldsymbol{\sigma}_{E,i} := \begin{bmatrix} z_i \\ f_{E,i} \end{bmatrix}, \quad \boldsymbol{\sigma}_{D,i} := \begin{bmatrix} v_i \\ f_{D,i} \end{bmatrix}, \quad \boldsymbol{\sigma}_i := \begin{bmatrix} z_i \\ v_i \\ f_i \end{bmatrix}. \tag{2.57}$$

According to the a priori process model (2.31), the force coordinates (f_E, f_D and f) are not estimated directly by the KF. They are nonlinear functions of the local random state vector $\mathbf{X}_C$, which is defined by (2.6) and (2.7). The calculation of the force coordinates has to be treated as the probabilistic transformation of $\mathbf{X}_C$ through these functions, since the estimated state vector is expected to be significantly corrupted by noise (as discussed in Section 2.2.4). Section 1.6.2 presented some flexible approximate solutions to this problem. However, since the particular transformation is based on functions, which are basic elements of the presented identification method that remain unchanged regardless of the system model, it is worth deriving an exact solution that is fast and accurate.

Let us start with the calculation of the probabilistic samples of eRFCs and dRFCs as a combined four-dimensional vector, denoted by $\mathbf{X}_{\mathrm{RFC}}$, defined by the vector function

$$\mathbf{X}_{\mathrm{RFC}} := \begin{bmatrix} Z \\ F_E \\ V \\ F_D \end{bmatrix} = \boldsymbol{\lambda}^{\mathrm{RFC}}(\mathbf{X}_C) = \begin{bmatrix} Z \\ S \cdot Z \\ V \\ D \cdot V \end{bmatrix}, \tag{2.58}$$

where capital notation indicates probability distributions of the related variables. The mean vector $\mathbf{x}_{\mathrm{RFC}}$ and the covariance matrix $\mathbf{P}_{\mathrm{RFC}}$ of the random vector variable $\mathbf{X}_{\mathrm{RFC}}$ can be expressed based on the Taylor series expansion of $\boldsymbol{\lambda}^{\mathrm{RFC}}$ around the mean value of $\mathbf{X}_C$, which is of the form

$$\boldsymbol{\lambda}^{\mathrm{RFC}}(\mathbf{X}_C) = \begin{bmatrix} z + (Z - z) \\ s \cdot z + s(Z - z) + (S - s)z + (S - s)(Z - z) \\ v + (V - v) \\ d \cdot v + d(V - v) + (D - d)v + (D - d)(V - v) \end{bmatrix}, \tag{2.59}$$

$$\mathbf{x}_{\mathrm{RFC}} = \mathrm{E}\left(\boldsymbol{\lambda}^{\mathrm{RFC}}(\mathbf{X}_C)\right), \tag{2.60}$$

$$\mathbf{P}_{\mathrm{RFC}} = \mathrm{E}([\boldsymbol{\lambda}^{\mathrm{RFC}}(\mathbf{X}_C) - \mathbf{x}_{\mathrm{RFC}}][\boldsymbol{\lambda}^{\mathrm{RFC}}(\mathbf{X}_C) - \mathbf{x}_{\mathrm{RFC}}]^{\mathrm{T}}). \tag{2.61}$$

In case of $\boldsymbol{\lambda}^{\mathrm{RFC}}$, the Taylor series expansion (2.59) gives an exact formula, since after the third term of the series all additional terms equal zero. The aim of the current derivation is to define $\mathbf{x}_{\mathrm{RFC}}$ and $\mathbf{P}_{\mathrm{RFC}}$ as functions of $\mathbf{x}_C$ and $\mathbf{P}_C$ to obtain explicit formulas for the RFC samples. Assuming that the local state vector is represented by a symmetric distribution and carrying out the algebraic operations defined by (2.60) and (2.61), the following exact formulas are derived:

$$\mathbf{x}_{\mathrm{RFC}} := \begin{bmatrix} \boldsymbol{\sigma}_E \\ \boldsymbol{\sigma}_D \end{bmatrix} = \begin{bmatrix} z \\ f_E \\ v \\ f_D \end{bmatrix} = \begin{bmatrix} z \\ s \cdot z + P_{ZS} \\ v \\ d \cdot v + P_{VD} \end{bmatrix}, \tag{2.62}$$

$$\mathbf{P}_{\mathrm{RFC}} := \begin{bmatrix} \mathbf{P}_{\sigma E} & \mathbf{P}_{\sigma E \sigma D} \\ \mathbf{P}_{\sigma E \sigma D}^{\mathrm{T}} & \mathbf{P}_{\sigma D} \end{bmatrix}, \tag{2.63}$$

$$\mathbf{P}_{\sigma E} = \begin{bmatrix} P_Z & sP_Z + zP_{ZS} \\ \cdots & s^2P_Z + 2zsP_{ZS} + z^2P_S + P_ZP_S + P_{ZS}^2 \end{bmatrix}, \tag{2.64}$$

$$\mathbf{P}_{\sigma D} = \begin{bmatrix} P_V & dP_V + vP_{VD} \\ \cdots & d^2P_V + 2vdP_{VD} + v^2P_D + P_VP_D + P_{VD}^2 \end{bmatrix}, \tag{2.65}$$

$$\mathbf{P}_{\sigma E \sigma D} = \begin{bmatrix} P_{ZV} & dP_{ZV} + vP_{ZD} \\ sP_{ZV} + zP_{VS} & sdP_{ZV} + vsP_{ZD} + zdP_{VS} + zvP_{SD} + P_{ZV}P_{SD} + P_{VS}P_{ZD} \end{bmatrix}, \tag{2.66}$$

where $\mathbf{P}_{\sigma E}$, $\mathbf{P}_{\sigma D}$ and $\mathbf{P}_{\sigma E\sigma D}$ denote the covariance matrices of $\boldsymbol{\sigma}_E$ and $\boldsymbol{\sigma}_D$ and their cross-covariance matrix respectively. The detailed derivation of these formulas is summarised in Appendix F. Notice that the mean values of the force coordinates in (2.62) are influenced by covariance values. Neglecting this fact (i.e. the deterministic calculation of the RFC samples) would produce a systematic error of the sample coordinates.

In order to reconstruct general RFSs (that can even describe coupled nonlinearities), formulas for the mean vector $\boldsymbol{\sigma}$ and the covariance matrix $\mathbf{P}_\sigma$ of the probabilistic RFS samples have to be derived as well. According to (1.15) the total restoring force f is the sum of the elastic and the dissipative restoring force components (f_E and f_D). Therefore, the probability distribution of $\boldsymbol{\sigma}$ can be defined as a function of $\mathbf{X}_{\mathrm{RFC}}$ in the form

$$\begin{bmatrix} Z \\ V \\ F \end{bmatrix} = \boldsymbol{\lambda}^{\mathrm{RFS}}(\mathbf{X}_{\mathrm{RFC}}) = \begin{bmatrix} Z \\ V \\ F_E + F_D \end{bmatrix}. \tag{2.67}$$

Since $\boldsymbol{\lambda}^{\mathrm{RFS}}$ is a linear function, the probabilistic transformation Φ of type LinT of the form (2.68) can be deployed to get the exact formulas (2.69) and (2.70).

$$[\boldsymbol{\sigma} \quad \mathbf{P}_\sigma \quad \cdots] = \Phi_{\boldsymbol{\lambda}^{\mathrm{RFS}}}^{\mathrm{LinT}}(\mathbf{x}_{\mathrm{RFC}}, \mathbf{P}_{\mathrm{RFC}}) \tag{2.68}$$

$$\boldsymbol{\sigma} = \begin{bmatrix} z \\ v \\ s \cdot z + d \cdot v + P_{ZS} + P_{VD} \end{bmatrix} \tag{2.69}$$

$$\mathbf{P}_{\sigma} = \begin{bmatrix} P_Z & P_{ZV} & sP_Z + zP_{ZS} + dP_{ZV} + vP_{ZD} \\ \cdots & P_V & sP_{ZV} + zP_{VS} + dP_V + vP_{VD} \\ \cdots & \cdots & P_{\mathrm{RFC},2,2} + 2 \cdot P_{\mathrm{RFC},2,4} + P_{\mathrm{RFC},4,4} \end{bmatrix} \tag{2.70}$$

The generated probabilistic sample datasets $\{\boldsymbol{\sigma}_E,\mathbf{P}_{\sigma E}\}_i$, $\{\boldsymbol{\sigma}_D,\mathbf{P}_{\sigma D}\}_i$ and $\{\boldsymbol{\sigma},\mathbf{P}_{\sigma}\}_i$ can now be used for the optimal fitting of the jth eRFC, dRFC and cRFS respectively.

2.3.2 Optimal nonparametric RFC fitting

The current section deals with the reconstruction of the noisefree eRFC and dRFC, i.e. $c_E(z)$ and $c_D(v)$, based on the noisy datasets $\{\boldsymbol{\sigma}_E,\mathbf{P}_{\sigma E}\}_i$ and $\{\boldsymbol{\sigma}_D,\mathbf{P}_{\sigma D}\}_i$ respectively that have been derived in Section 2.3.1. These two curves define an additive RFS according to (1.9). They can be calculated separately using the same algorithm. Therefore, the technique that is derived in the following for the elastic curve c_E can straightforwardly be applied to the dissipative curve c_D as well. The dataset $\{\boldsymbol{\sigma}_E,\mathbf{P}_{\sigma E}\}_i$ can be treated as a noisy virtual measurement of the real eRFC in the form

$$\boldsymbol{\sigma}_{E,i} := \begin{bmatrix} z_i \\ f_{E,i} \end{bmatrix} = \begin{bmatrix} z_{\mathrm{true},i} \\ c_E(z_{\mathrm{true},i}) \end{bmatrix} + \mathbf{r}_{\sigma E,i}\,, \tag{2.71}$$

$$\mathbf{r}_{\sigma E} \sim (\mathrm{E}(\mathbf{r}_{\sigma E}), \mathbf{P}_{\sigma E})\,, \tag{2.72}$$

where $\mathbf{r}_{\sigma E}$ denotes the random noise process that is corrupting the virtual measurement. Although the formulas derived in Section 2.3.1 ensure the unbiased transformation of the estimated state vector $\mathbf{x}_C$ into eRFC samples $\boldsymbol{\sigma}_E$, the noise process $\mathbf{r}_{\sigma E}$ still includes bias caused by the estimation error of the KF, i.e. by the deviation of $\mathbf{x}_C$ from its true values. Though the tuning strategy presented in Section 2.2.4 aims to minimise this error (for the price of increased noisiness of the state estimates), the results of the KF will not be perfect, therefore, the bias of $\mathbf{r}_{\sigma E}$, i.e. $\mathrm{E}(\mathbf{r}_{\sigma E})$, will not be zero. This bias is however unknown, hence, it has to be neglected in the curve fitting algorithm. Notice that (in contrast to the usual noise vector of real measurements) $\mathbf{r}_{\sigma E}$ is essentially correlated, i.e.

$\mathbf{P}_{\sigma E}$ is not a diagonal matrix, since all its coordinates originate from the same noise process, namely that of the real measurement vector $\mathbf{y}$. Therefore, the curve fitting problem can be formulated as the minimisation of the Weighted Total Squares error e^{WTS} defined as

$$\mathbf{e}_{cE,i} = \boldsymbol{\sigma}_{E,i} - \begin{bmatrix} \hat{c}_E^{-1}(f_{E,i}) \\ \hat{c}_E(z_i) \end{bmatrix}, \tag{2.73}$$

$$e_{cE}^{\text{WTS}} = \sum_{i=1}^{n_S} \mathbf{e}_{cE,i}{}^{\text{T}} \mathbf{P}_{\sigma E,i}^{-1} \mathbf{e}_{cE,i}\,, \tag{2.74}$$

where $\hat{c}_E$ and $\hat{c}_E^{-1}$ denote the estimate of the true RFC, i.e. c_E, and its inverse function respectively. The function $\hat{c}_E$ is an optimal estimate in Weighted Total Least Squares (WTLS) sense if it is the minimiser of e_{cE}^{WTS}. An RFC of a physical system is per definition always a unique function. This however not necessarily holds for its inverse function. Hence, the calculation of the total error vector $\mathbf{e}_{cE}$ can easily become a nontrivial task, which is however efficiently handled by the technique that is proposed in the current section.

In the present thesis a nonparametric regression is applied to the WTLS fitting problem of the RFCs in order to keep the developed identification method completely nonparametric. To do so, an alternative approach is proposed that has two advantages compared to existing solutions that have been summarised in Section 1.7. First, it is suitable for general correlated noise, i.e. for non-diagonal $\mathbf{P}_{\sigma E}$. Second, compared to other nonparametric algorithms, such as the LPR (q.v. Section 1.7), it is a computationally cheaper algorithm, because it is based on a smaller number of calls per dataset sample. It is not a general WTLS approach, because it crucially implies that the dataset $\{\boldsymbol{\sigma}_E,\mathbf{P}_{\sigma E}\}_i$ is given in form of time signals. This however always holds in case of the RFC samples, therefore, it does not represent any restriction to the presented identification method.

Our aim is to generate a smoothed (noisefree) version of this observed dataset, denoted by $\boldsymbol{\sigma}_{E,i}^{\mathrm{S}}$, which represents the optimal estimate of the eRFC, i.e. $\hat{c}_E(z)$, that is a minimiser of e_{cE}^{WTS}. To do so, the task is considered as a two-dimensional target tracking problem, where the non-smooth dataset is considered as the noisy measurement of a target's movement on the plane of the eRFC coordinates from one end of the true curve c_E to the other end, i.e. from $\min(z_{\mathrm{true},i})$ to $\max(z_{\mathrm{true},i})$, with a priori known correlated measurement error $\mathbf{P}_{\sigma E}$. In order to make the sample dataset "look like" such a measurement, the samples are sorted in ascending order of the coordinate z (i.e. the deformation of the eRFC). This idea is proposed in (Young, 2000) as a part of the method of State Dependant Parameters (SDP). The smoothed path of the imaginary target is then estimated using a recursive 0th order random walk estimator (Young, 2000), also known as the α-filter (Simon, 2006). This is actually a special case of the EKF with two state variables (z and f_E), therefore, the formulas presented in Section 1.6.2 and Section 1.6.3 can directly be applied after defining the process function and the measurement function, i.e. **a** and **h**, as

$$\mathbf{a}^{cE}(z, f_E) := \begin{bmatrix} 0 \\ 0 \end{bmatrix}, \qquad \mathbf{h}^{cE}(z, f_E) := \begin{bmatrix} z \\ f_E \end{bmatrix}. \tag{2.75}$$

The sequence of recursion is now defined by the spatial coordinate z instead of time. The measurement covariance matrix **R** is replaced by $\mathbf{P}_{\sigma E,i}$ that in this case has different values at each recursion step. Due to this, the estimation quality feedback of the KF is directly taken into account in the fitting process, which provides a higher weighting of reliable samples. The choice of the discrete time (precisely speaking "discrete spatial step") process covariance matrix **Q**, here denoted by $\mathbf{Q}^{cE}$, is discussed in detail later on in this section. The a posteriori state estimates of this modified KF form the smoothed dataset $\boldsymbol{\sigma}_{E,i}^{\mathrm{S}}$. The a posteriori covariance matrix of the state estimation error, denoted in this particular case by $\mathbf{P}_{\sigma E,i}^{\mathrm{S}}$, which is a standard result of the KF, provides a quantification of the estimated eRFC's uncertainty. It is generated as a nonparametric function of the deformation coordinate z and provides a useful feedback on the estimation quality, which is given separately for the two coordinates (z and f_E) of the eRFCs nodes. This satisfies the corresponding requirement from Section 1.1.

Notice that the two ends of the spatial recursion have been defined previously as $\min(z_{\mathrm{true},i})$ and $\max(z_{\mathrm{true},i})$, although the true values of z are not known prior to the curve fitting. This is a key point of the proposed algorithm. Actually, in order to achieve a WTLS fit, not only the two ends of the recursion but the whole sorting of the dataset and the subsequential recursion has to be carried out over the true values of z. This can not be realised perfectly, but a sufficiently accurate approximation of z_{true} can be given by the Central Moving Average (q.v. Appendix E) of the noisy z, denoted by z^{CMA}. Thereby some level of smoothness of the noisefree z_{true} is assumed. This is the point, where the current approach implies the observed dataset to be given by realistic time signals, which automatically ensures some natural smoothness. Due to this solution, the procedure has a slightly iterative manner, since a first guess of the smoothed dataset's first coordinates is generated in order to initiate the main fitting algorithm. The procedure has been implemented without additional iteration cycles of the main fitting step, which is (as the following examples show) sufficient to achieve good results.

To illustrate the necessity of the discussed kind of sorting, let us take a synthetic example of a highly nonlinear eRFC, where the time series $\{\boldsymbol{\sigma}_E,\mathbf{P}_{\sigma E}\}_i$ is directly generated and corrupted by artificial correlated noise. This avoids the possible effect of the previously mentioned bias of the results of the first KF run, i.e. it ensures $\mathrm{E}(\mathbf{r}_{\sigma E})=\mathbf{0}$. The true and the noisy values of the two eRFC coordinates are depicted in Figure 2.11, which shows a 0.01 s time segment of the dataset that has been observed with 10 kHz sampling rate over 1 s. The true deformation is defined as a two-component (173 Hz and 633 Hz) harmonic oscillation. Its observations are corrupted by stationary zero-mean Gaussian white noise with a covariance matrix of

$$\mathbf{P}_{\sigma E,i} := \begin{bmatrix} 5 \cdot 10^{-12} & 1.5 \cdot 10^{-5} \\ 1.5 \cdot 10^{-5} & 75 \end{bmatrix}. \qquad \textbf{(2.76)}$$

This represents a higher noise level than what can be expected under realistic conditions (q.v. Chapter 3 and 5), which is ideal as an extreme test of the presented technique. The five-point CMA of the observed signals (denoted by CMA5) is shown in Figure 2.11 as well to illustrate that (though z^{CMA} is suitable as recursion coordinate) these signals could not be directly used as noisefree sample coordinates due to their deviation from the true values.

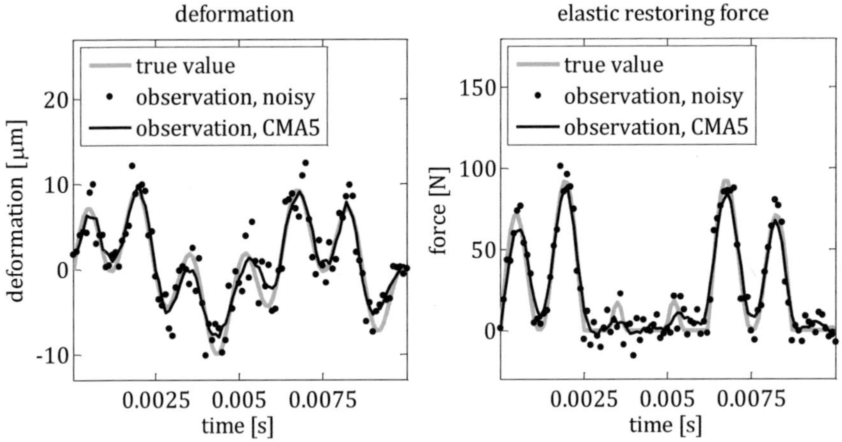

Figure 2.11: Synthetic (directly generated) example of the sample dataset of a highly nonlinear eRFC generated at 10 kHz observation sampling rate and corrupted by correlated Gaussian white noise.

The effect of the proper choice of the recursion coordinate becomes clearer by taking a look at Figure 2.12 that illustrates the simple sorting over z in comparison with the enhanced sorting over z^{CMA} based on the defined numerical example. Notice that the observed samples represent a much better image of the true signal's statistics in case of the enhanced sorting, which is a crucial requirement for the target tracking approach described above.

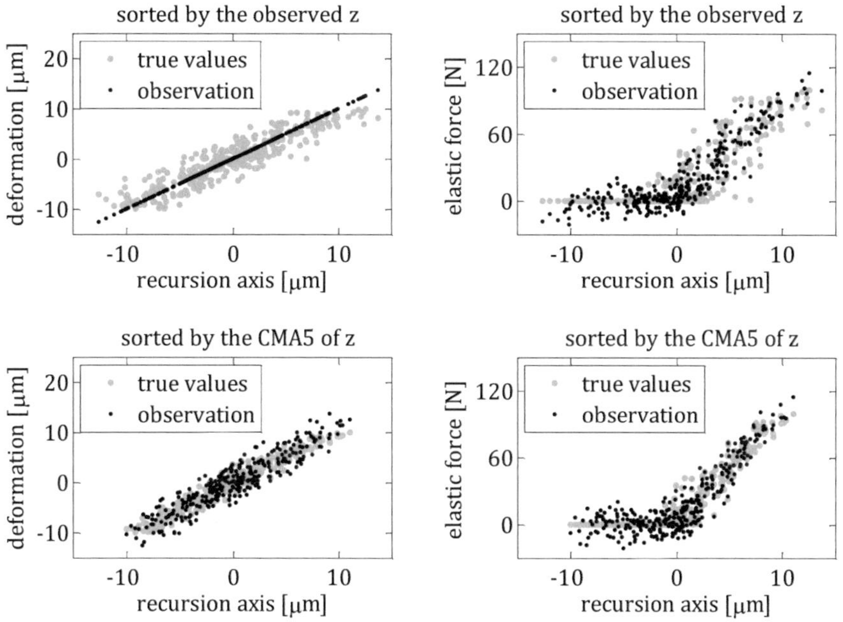

Figure 2.12: Sorted samples of a highly nonlinear eRFC based on synthetic correlated noisy observations. Sorting over the observed deformation (top) is compared to the sorting over the CMA5 moving averaged deformation (bottom).

Now that the samples are prepared for the modified KF (q.v. (2.75)), the last missing element of the technique is the proper choice of the process covariance matrix $\mathbf{Q}^{cE}$, that controls the smoothness of the identified RFC. As follows, a simple formula is suggested for choosing its values. To do so, let us first consider a one dimensional tracking problem, where only the force coordinate f_E is observed. The resulting formula is then extended to the two dimensional case at the end. The spatial recursion axis z^{CMA} is treated as if it was time regarding its nomenclature throughout this short derivation in order to keep the context clearer. This means that it is denoted by t and derivatives with respect to it are indicated by the "dot" symbol. Let us describe the true characteristic of the elastic force by an average slope over the recursion axis, denoted by $\Delta f_E/\Delta t$. This can be determined as the slope of the rectangle's diagonal that envelopes the observed samples in the $\{z^{\mathrm{CMA}},f_E\}$ plane.

Given a constant continuous-time process covariance of Q^{cE}_{cont} the dynamics of the smoothed elastic force, denoted by f^S_E, are approximately governed by the differential equation (2.77) except for a short initial transient phase (Simon, 2006). The average of the continuous-time measurement covariance is thereby denoted by R^{cE}_{cont}. Equation (2.77) can be transformed into the error equation (2.79) by defining the normed instantaneous curve estimation error e_c as (2.78). By reformulating the stationary solution of the error equation, given by (2.80), the value of Q^{cE}_{cont} is found as a function of the normed stationary curve estimation error, denoted by $e_{c,\infty}$, given as (2.81).

$$\dot{f}^S_E = \sqrt{\frac{Q^{cE}_{\mathrm{cont}}}{R^{cE}_{\mathrm{cont}}}}\left(\frac{\Delta f_E}{\Delta t}t - f^S_E\right) \tag{2.77}$$

$$e_c := \frac{t}{\Delta t} - \frac{f^S_E}{\Delta f_E} \tag{2.78}$$

$$\dot{e}_c = \frac{1}{\Delta t} - \sqrt{\frac{Q^{cE}_{\mathrm{cont}}}{R^{cE}_{\mathrm{cont}}}}\, e_c \tag{2.79}$$

$$e_{c,\infty} = \frac{1}{\Delta t}\sqrt{\frac{R^{cE}_{cont}}{Q^{cE}_{cont}}} \tag{2.80}$$

$$Q^{cE}_{cont} = \frac{R^{cE}_{cont}}{e^2_{c,\infty} \cdot \Delta t^2} \tag{2.81}$$

Using the rules (2.82) (Simon, 2006) for the transformation of the continuous-time variances into their discrete-time counterparts, denoted by Q^{cE} and R^{cE}, the formula (2.81) becomes (2.83), where Q^{cE} is a function of the recursion step size dt and is therefore not constant over the recursion.

$$Q^{cE}_{\mathrm{cont}} = Q^{cE}/dt\,, \qquad R^{cE}_{\mathrm{cont}} = R^{cE}dt \tag{2.82}$$

$$Q_i^{cE} = \frac{dt_i^2 \cdot R^{cE}}{e_{c,\infty}^2 \cdot \Delta t^2} \quad (2.83)$$

Extending this result to the correlated two dimensional case of the RFC fitting problem and switching back to the notation z^{CMA} instead of t, the final formula for the tuning matrix $\mathbf{Q}^{cE}$ is given for the ith recursion step as (2.84), which provides a uniform smoothing of the identified RFC.

$$\mathbf{Q}_i^{cE} := \frac{\left(dz_i^{\mathrm{CMA}}\right)^2 \cdot \mathrm{E}(\mathbf{P}_{\sigma E})}{e_{c,\infty}^2 \cdot \left(\Delta z_i^{\mathrm{CMA}}\right)^2} \quad (2.84)$$

Accordingly, the proposed nonparametric curve fitting algorithm is tuned by the averaging window size of the CMA and the normed error $e_{c,\infty}$. The optimisation of these two tuning factors regarding the identified RFC's *eRMS* error based on a group of curve examples (q.v. Figure 2.15) yielded a CMA window size of 5 samples (i.e. CMA5) in combination with $e_{c,\infty}$=0.01 as the best choice. These settings are used for RFC fitting throughout the rest of the current thesis. The smoothed dataset of the nonlinear eRFC from the previously defined numerical example is depicted in Figure 2.13 showing the significant increase of result quality due to the choice of z^{CMA5} instead of z as the coordinate of recursion. The presented procedure obviously does not reduce the number of samples. However, since a noisefree dataset has been identified, the number of characteristic points can now easily be reduced by resampling without significant loss of information in order to generate compact lookup tables that are practical for simulation purposes or other types of system analyses.

In order to indicate the local uncertainty of the smoothed dataset, the a posteriori variance $\mathbf{P}_{\sigma E}^{\mathrm{S}}$ of the eRFC samples can be used to generate lower and upper uncertainty curves given by the datasets $\boldsymbol{\rho}_{E,i}^{\mathrm{L}}$ and $\boldsymbol{\rho}_{E,i}^{\mathrm{U}}$ respectively as

$$\boldsymbol{\rho}_{E,i}^{\mathrm{L}} := \boldsymbol{\sigma}_{E,i}^{\mathrm{S}} + p_{\mathrm{plot}} \begin{bmatrix} +\sqrt{P_{\sigma E,1,1}^{\mathrm{S}}} \\ -\sqrt{P_{\sigma E,2,2}^{\mathrm{S}}} \end{bmatrix}_i , \quad \textbf{(2.85)}$$

$$\boldsymbol{\rho}_{E,i}^{\mathrm{U}} := \boldsymbol{\sigma}_{E,i}^{\mathrm{S}} + p_{\mathrm{plot}} \begin{bmatrix} -\sqrt{P_{\sigma E,1,1}^{\mathrm{S}}} \\ +\sqrt{P_{\sigma E,2,2}^{\mathrm{S}}} \end{bmatrix}_i , \quad \textbf{(2.86)}$$

where $P_{\sigma E,j,j}^{\mathrm{S}}$ denote corresponding diagonal elements of $\mathbf{P}_{\sigma E}^{\mathrm{S}}$ and p_{plot} serves as an optical tuning factor for the visualisation of the uncertainty curves together with the fitted eRFC. Notice that the formulas (2.85) and (2.86) are given for RFCs with positive slope. The signs of the variance dependent offset terms have to be switched in case of negative slope. For the special case of non-monotonic RFCs a more complex formulation is necessary.

As the final step of the presented procedure the unreliable samples, which typically occur at the two ends of the smoothed dataset, are discarded from the results. The introduced uncertainty curves are perfectly suitable for this purpose, which can be seen in Figure 2.14, where an illustrated summary of the whole curve fitting algorithm is given from the noisy RFC samples to the final smoothed, trimmed and resampled RFC (based on the previously defined numerical example).

Further numerical examples of four different RFCs are depicted in Figure 2.15, which aims to prove the flexibility of the proposed technique.

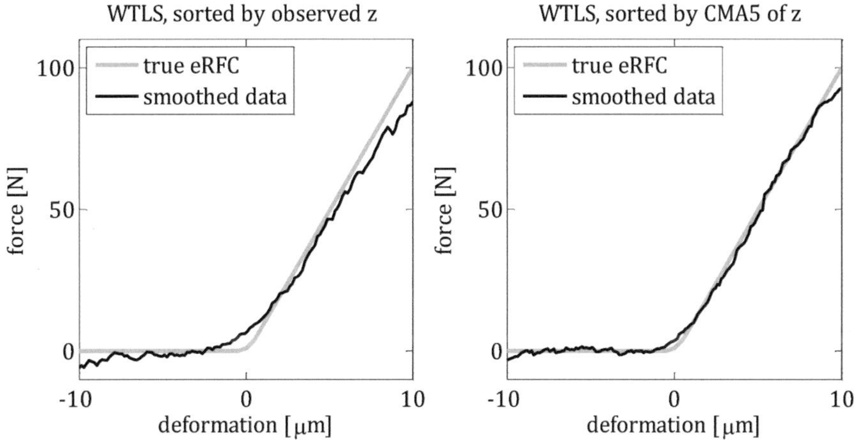

Figure 2.13: The true curve of a highly nonlinear eRFC is compared to the WTLS smoothed samples based on synthetic correlated noisy observations. Sorting over the observed deformation (left) is compared to the sorting over the CMA5 moving averaged deformation (right).

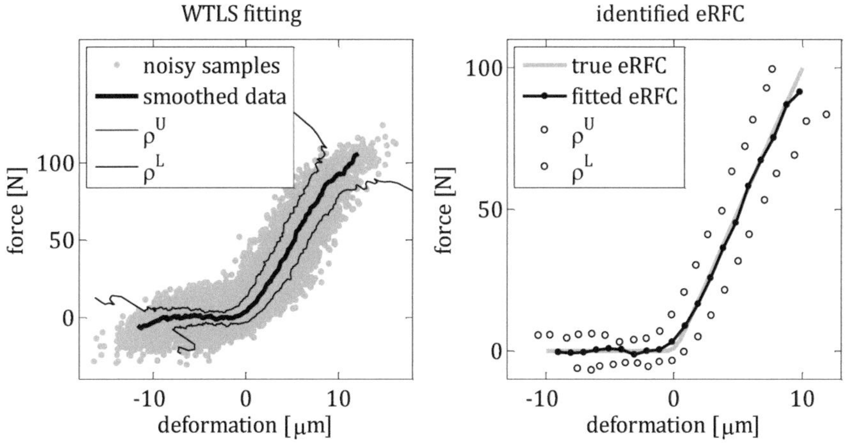

Figure 2.14: Illustrated summary of the proposed nonparametric WTLS curve fitting algorithm based on the synthetic correlated noisy samples of a highly nonlinear eRFC.

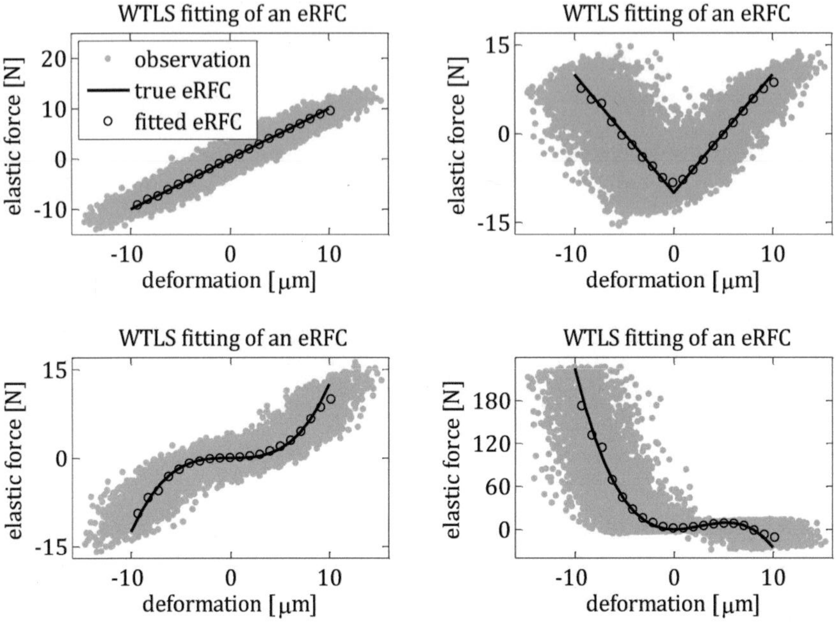

Figure 2.15: Different RFC fitting examples carried out using the proposed nonparametric WTLS approach based on synthetic correlated noisy observations.

2.3.3 Optimal nonparametric RFS fitting

The current section deals with the direct reconstruction of a noisefree RFS, i.e. $c(z,v)$, based on the noisy dataset $\{\boldsymbol{\sigma},\mathbf{P}_\sigma\}_i$ that has been derived in Section 2.3.1. This surface defines a general coupled RFS (cRFS) that is useful in situations, where the assumption of an additive nonlinearity according to (1.9) is not justified. The fitting of an RFS follows the same logic as the fitting of the RFCs, therefore, at many points of the current section it is referred to Section 2.3.2 instead of repeating the same reasoning. The dataset $\{\boldsymbol{\sigma},\mathbf{P}_\sigma\}_i$ can be treated as a noisy virtual measurement of the real RFS in the form

$$\boldsymbol{\sigma}_i := \begin{bmatrix} z_i \\ v_i \\ f_i \end{bmatrix} = \begin{bmatrix} z_{\text{true},i} \\ v_{\text{true},i} \\ c(z_{\text{true},i}, v_{\text{true},i}) \end{bmatrix} + \mathbf{r}_{\sigma,i}\,, \tag{2.87}$$

$$\mathbf{r}_\sigma \sim (\mathrm{E}(\mathbf{r}_\sigma), \mathbf{P}_\sigma)\,, \tag{2.88}$$

where $\mathbf{r}_\sigma$ denotes the random noise process that is corrupting the virtual measurement. As discussed in Section 2.3.2, the results of the augmented KF (q.v. Section 2.2) will not be perfect, therefore, the bias of $\mathbf{r}_\sigma$, i.e. $\mathrm{E}(\mathbf{r}_\sigma)$, will not be zero. This bias however has to be neglected in the fitting algorithm, since it is of unknown nature. As in the case of the RFCs, $\mathbf{r}_\sigma$ is essentially correlated, i.e. $\mathbf{P}_\sigma$ is not a diagonal matrix. Therefore, the surface fitting problem can be formulated as the minimisation of the Weighted Total Squares error e^{WTS} defined as

$$\mathbf{e}_{c,i} = \boldsymbol{\sigma}_i - \begin{bmatrix} \hat{c}^{v,f}(v_i, f_i) \\ \hat{c}^{z,f}(z_i, f_i) \\ \hat{c}(z_i, v_i) \end{bmatrix}, \tag{2.89}$$

$$e_c^{\mathrm{WTS}} = \sum_{i=1}^{n_s} \mathbf{e}_{c,i}{}^{\mathrm{T}} \mathbf{P}_{\sigma,i}^{-1} \mathbf{e}_{c,i}\,, \tag{2.90}$$

where $\hat{c}$, $\hat{c}^{z,f}$ and $\hat{c}^{v,f}$ denote the estimate of the true RFS, i.e. $c(z,v)$, and its two reformulated versions that accommodate $\{z_i,f_i\}$ and $\{v_i,f_i\}$ as independent variable pairs respectively. The function $\hat{c}$ is an optimal estimate in Weighted Total Least Squares (WTLS) sense if it is the minimiser of e_c^{WTS}. The values of the reformulated functions $\hat{c}^{z,f}$ and $\hat{c}^{v,f}$ are not necessarily unique in general, hence, the direct calculation of the total error vector $\mathbf{e}_c$ can cause difficulties. This problem is efficiently handled by the technique that is proposed in the following.

The brief literature overview in Section 1.7 discusses state of the art fitting approaches that can handle surface fitting as well. One can conclude in compact form that there are powerful nonparametric surface fitting solutions that are based on Local Polynomial Regression (Fox, 2002), (Zhang, et al., 2013), which however do not consider samples corrupted by fully correlated noise. Unfortunately the idea of considering the WTLS regression problem as target tracking, as it is applied for curve fitting in Section 2.3.2, can not be applied for surfaces, since they do not represent a path, therefore, the definition of an appropriate recursion coordinate is not possible. Accordingly, an alternative strategy has been implemented in the present thesis that is also nonparametric in order to completely avoid functional assumptions in the developed identification method. The approach has the following two-step structure:

- Step 1 is a local WTLS point fitting, where the observed RFS samples are sorted into groups based on a predefined segmenting of the {*z*,*v*} plane. Each of these sample groups are replaced by one fitted point, denoted by $\boldsymbol{\sigma}_j^{\mathrm{WTLS}}$, which represents their mean value in the {*z*,*v*,*f*} space in WTLS sense (q.v. Figure 2.16 left and Figure 2.17 left). This simple case of WTLS fitting has an explicit solution for correlated noise.

- Step 2 is a quadratic Local Polynomial Regression (LPR) of the WTLS points, i.e. of the results of step 1, over a predefined uniform grid of the {*z*,*v*} plane. It provides the final identified RFS in form of a lookup table of its smoothed nodes, denoted by $\boldsymbol{\sigma}_k^{\mathrm{S}}$, (q.v. Figure 2.16 right and Figure 2.17 right). This additional regression is necessary due to two reasons. First, in step 1 a fitting in all three dimensions of the RFS is carried out, therefore, the resulting points are not arranged on a uniform grid in the {*z*,*v*} plane, which is disadvantageous for the further use of the RFS as a lookup table. Second, in case of insufficient number of observed samples further smoothing of the WTLS points is reasonable, due to the slow statistical convergence of nonparametric surface fitting (q.v. Section 3.4.2).

A detailed description of these two steps is presented in the following including the illustration of the approach based on virtually generated examples.

Step 1: local WTLS point fitting

The local point fitting generates a reduced number of samples, denoted by $\boldsymbol{\sigma}_j^{\text{WTLS}}$, which are calculated in WTLS sense. The index *j* denotes the *j*th segment of the {*z*,*v*} plane that includes the *j*th group of the original samples, each of which are replaced by one fitted point (q.v. Figure 2.16 left). In order to create a better image of the original samples' statistics, the sorting of the samples into the predefined segments of the {*z*,*v*} plane is carried out based on their CMA5 averaged coordinates (z^{CMA5} and v^{CMA5}). This explains, why the *j*th group of samples in Figure 2.16 (left) is not bounded by a rectangle in the {*z*,*v*} plane. This solution is driven by the same logic as in Section 2.3.2 in case of the sorting of RFC samples. The reason of applying the first step in this form is that the special case of fitting a single point in WTLS sense has an exact solution, which according to (Simon, 2006) can be formulated in a recursive algorithm as:

$$\boldsymbol{\sigma}_{j,1}^{\text{WTLS}} := \boldsymbol{\sigma}_1 , \qquad \mathbf{P}_{\sigma,j,1}^{\text{WTLS}} := \mathbf{P}_{\sigma,1} \tag{2.91}$$

$$\mathbf{K}_{i+1}^{\text{WTLS}} := \mathbf{P}_{\sigma,j,i}^{\text{WTLS}} \left(\mathbf{P}_{\sigma,j,i}^{\text{WTLS}} + \mathbf{P}_{\sigma,i+1} \right)^{-1} \tag{2.92}$$

$$\mathbf{P}_{\sigma,j,i+1}^{\text{WTLS}} := \left(\mathbf{I} - \mathbf{K}_{i+1}^{\text{WTLS}} \right) \mathbf{P}_{\sigma,j,i}^{\text{WTLS}} \left(\mathbf{I} - \mathbf{K}_{i+1}^{\text{WTLS}} \right)^{\text{T}} + \mathbf{K}_{i+1}^{\text{WTLS}} \mathbf{P}_{\sigma,i+1} \left(\mathbf{K}_{i+1}^{\text{WTLS}} \right)^{\text{T}} \tag{2.93}$$

$$\boldsymbol{\sigma}_{j,i+1}^{\text{WTLS}} := \boldsymbol{\sigma}_{j,i}^{\text{WTLS}} + \mathbf{K}_{i+1}^{\text{WTLS}} \left(\boldsymbol{\sigma}_{i+1} - \boldsymbol{\sigma}_{j,i}^{\text{WTLS}} \right) \tag{2.94}$$

The *j*th recursion is carried out for $i \in \{1,\dots,n_j-1\}$, where the number of noisy samples in the *j*th group is denoted by n_j. The last values of the *j*th recursion represent the *j*th fitted WTLS point. These points build the dataset $\{\boldsymbol{\sigma}_j^{\text{WTLS}}, \mathbf{P}_{\sigma,j}^{\text{WTLS}}\}$, which forms the result of the fitting algorithm's first step. The reason of implementing this recursive formulation is that the also available explicit algorithm turned out to be computationally expensive due to the inversion of matrices of size $3n_j \times 3n_j$. The WTLS point fitting is illustrated in Figure 2.17 (left) based on the synthetic correlated noisy samples of a paraboloid surface. The observations are based on the two-component harmonic deformation time series z from the numerical example in Section 2.3.2 and its analytical time derivative as the rate of deformation v. The signals are corrupted by zero-mean Gaussian white noise with a covariance matrix of

$$\mathbf{P}_{\sigma,i} := \begin{bmatrix} 10^{-12} & 10^{-10} & 10^{-6} \\ 10^{-10} & 10^{-6} & 10^{-4} \\ 10^{-6} & 10^{-4} & 30 \end{bmatrix}. \tag{2.95}$$

Step 2: Local Polynomial Regression

During the present study it turned out that the dataset $\boldsymbol{\sigma}_j^{\mathrm{WTLS}}$ can not be directly used as a nonparametric RFS due to its insufficient smoothness. The reason for this originates from the statistical aspects that have been discussed in Section 2.2.1, which leads to a slow statistical convergence of nonparametric surface fitting (q.v. Section 3.4.2). To overcome this problem, the current approach is extended by a quadratic Local Polynomial Regression (LPR) in standard Weighted Least Squares (WLS) sense in order to achieve the smoothed and uniformly resampled set of surface points, denoted by $\boldsymbol{\sigma}_k^{\mathrm{S}}$. These form the end result of the RFS fitting in form of a lookup table. To do so, the desired resolution of the result is defined by generating a uniform grid of size $\sqrt{n_z}\times\sqrt{n_z}$ in the $\{z,v\}$ plane that is given by a set of coordinate pairs as

$$\mathcal{Z} := \begin{bmatrix} z_1 & \cdots & z_{n_z} \\ v_1 & \cdots & v_{n_z} \end{bmatrix}_{2\times n_z}^{\mathrm{T}}, \tag{2.96}$$

to which the n_z smoothed force coordinates, denoted by f^{S}, are to be determined by the fitting algorithm. Based on the theory presented in (Fox, 2002) the WLS quadratic LPR can be formulated as follows:

$$\mathbf{R}_{p,k} = \mathrm{diag}\left(\begin{bmatrix} P_{\sigma,1,f,f}^{\mathrm{WTLS}} & \cdots & P_{\sigma,n_k,f,f}^{\mathrm{WTLS}} \end{bmatrix}_{1\times n_k}\right) \tag{2.97}$$

$$\mathbf{H}_{p,k} = \begin{bmatrix} 1 & \sigma_{1,z}^{\mathrm{WTLS}} & \sigma_{1,v}^{\mathrm{WTLS}} & \sigma_{1,z}^{\mathrm{WTLS}}\sigma_{1,v}^{\mathrm{WTLS}} & \left(\sigma_{1,z}^{\mathrm{WTLS}}\right)^2 & \left(\sigma_{1,v}^{\mathrm{WTLS}}\right)^2 \\ & & & \vdots & & \\ 1 & \sigma_{n_k,z}^{\mathrm{WTLS}} & \sigma_{n_k,v}^{\mathrm{WTLS}} & \sigma_{n_k,z}^{\mathrm{WTLS}}\sigma_{n_k,v}^{\mathrm{WTLS}} & \left(\sigma_{n_k,z}^{\mathrm{WTLS}}\right)^2 & \left(\sigma_{n_k,v}^{\mathrm{WTLS}}\right)^2 \end{bmatrix}_{n_k \times 6} \tag{2.98}$$

$$\mathbf{P}_{\sigma,k}^{p} = \left(\mathbf{H}_{p,k}^{\mathrm{T}}\mathbf{R}_{p,k}^{-1}\mathbf{H}_{p,k}\right)^{-1} \tag{2.99}$$

$$\mathbf{p}_{\sigma,k} = \mathbf{P}_{\sigma,k}^{p}\mathbf{H}_{p,k}^{\mathrm{T}}\mathbf{R}_{p,k}^{-1}\left[\sigma_{1,f}^{\mathrm{WTLS}} \quad \cdots \quad \sigma_{n_k,f}^{\mathrm{WTLS}}\right]^{\mathrm{T}} \tag{2.100}$$

Thereby the number of WTLS points in the kth segment (q.v. Figure 2.16 right) is denoted by n_k, and the specific elements of vectors and matrices have been indexed using $\{z,v,f\}$ instead of $\{1,2,3\}$ in order to keep the notation clearer. The dataset $\{\mathbf{p}_{\sigma,k}, \mathbf{P}_{\sigma,k}^{p}\}$ represents the parameters of the kth locally fitted quadratic polynomial surface. These are then transformed into the locally smoothed RFS points over the predefined grid as follows:

$$\mathbf{H}_{\sigma,k} = \begin{bmatrix} 1 & \mathcal{Z}_{k,z} & \mathcal{Z}_{k,v} & \mathcal{Z}_{k,z}\mathcal{Z}_{k,v} & \mathcal{Z}_{k,z}^2 & \mathcal{Z}_{k,v}^2 \end{bmatrix}_{1\times 6} \tag{2.101}$$

$$\boldsymbol{\sigma}_k^{\mathrm{S}} = [\mathcal{Z}_{k,z} \quad \mathcal{Z}_{k,z} \quad \mathbf{H}_{\sigma,k}\mathbf{p}_{\sigma,k}]^{\mathrm{T}} \tag{2.102}$$

$$P_{\sigma,k}^{\mathrm{S}} = \mathbf{H}_{\sigma,k}\mathbf{P}_{\sigma,k}^{p}\mathbf{H}_{\sigma,k}^{\mathrm{T}} \tag{2.103}$$

This second step of the procedure is illustrated in Figure 2.17 (right) for the previously introduced paraboloid example. The dataset $\{\boldsymbol{\sigma}^{\mathrm{S}}, P_{\sigma}^{\mathrm{S}}\}_k$ forms the final nonparametric result of the presented approach.

The variance values P_{σ}^{S} provide useful information about the uncertainty of the fitted RFS's nodes, i.e. $\boldsymbol{\sigma}^{\mathrm{S}}$. Plotted as a function over z and v, they can be used as a quality feedback of the identified surface. This is illustrated on two nonlinear RFS examples in Figure 2.18 that have been generated based on synthetic correlated noisy samples.

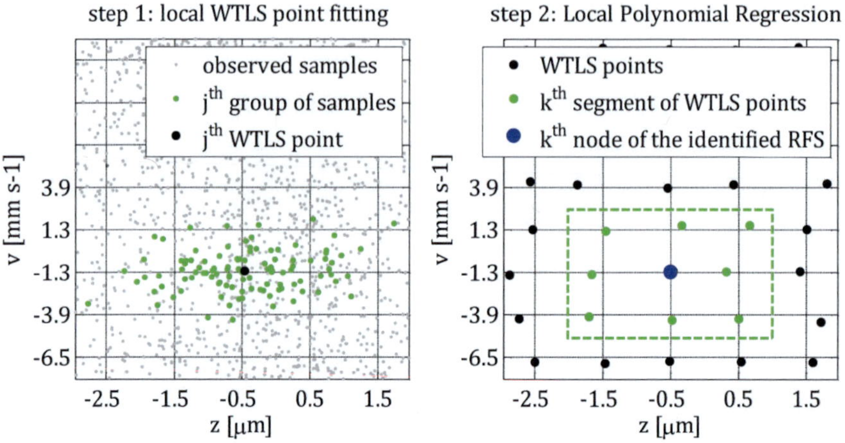

Figure 2.16: Illustration of the two steps of the presented nonparametric surface fitting approach in the {*z*,*v*} plane. Intermediate samples are generated using local WTLS point fitting (left). The intermediate samples are then used in a subsequential quadratic Local Polynomial Regression (right).

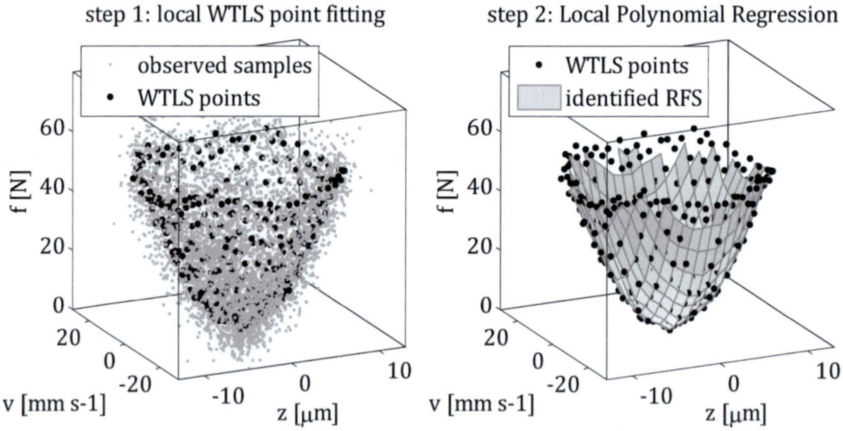

Figure 2.17: The two steps of the presented nonparametric surface fitting approach based on the synthetic correlated noisy samples of a paraboloid RFS. Intermediate samples are generated using local WTLS point fitting (left). The intermediate samples are then used in a subsequential quadratic Local Polynomial Regression (right).

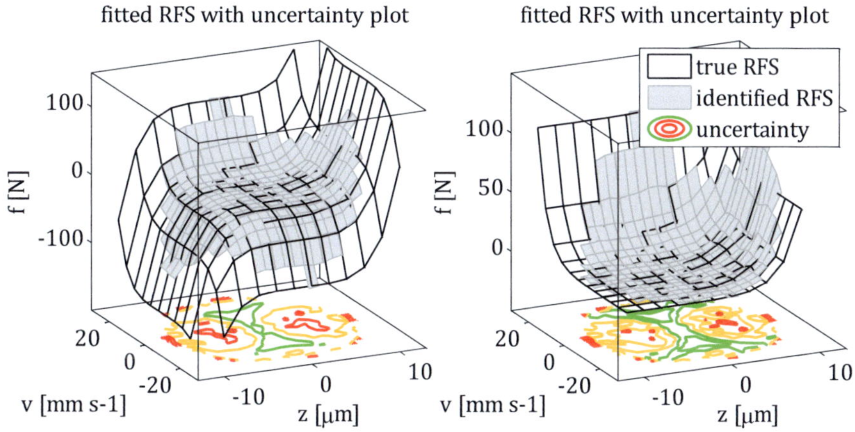

Figure 2.18: Comparison of the true and the identified surfaces of two nonlinear RFSs. Fitting was carried out using the presented two-step nonparametric approach based on synthetic correlated noisy samples.

Notice that the three coordinates (z, v and f) of typical RFSs of mechanical engineering systems (with high characteristic frequencies) strongly differ in their order of magnitude, which would lead to bad conditioning of the surface fitting formulas. To account for this, an appropriate scaling of the dataset $\{\boldsymbol{\sigma},\mathbf{P}_{\sigma}\}_i$ is carried out prior to the fitting of the RFS (using a proper probabilistic transformation of type LinT from Section 1.6.2). The algorithm is however presented without scaling in the current section, in order to keep the notations as simple as possible.

2.3.4 Conclusion of the proposed fitting approach

In Section 2.3 a nonparametric (approximately) WTLS fitting approach has been proposed that allows the reconstruction of arbitrary noisefree RFCs and RFSs based on the results of the augmented KF given as (2.56). In accordance with the requirements formulated in Section 1.1 the presented technique exhibits the following important properties:

- The fitted curves (RFCs) and surfaces (RFSs) are given nonparametrically in form of lookup tables. No assumptions about their functional form are required.
- Throughout the definition of the procedure great attention was paid to keep the algorithm computationally as simple as possible.
- Additional feedback is given about the quality (uncertainty) of the fitted results in form of uncertainty plots (q.v. Figure 2.14 and Figure 2.18).

Since state of the art modelling is mostly carried out virtually on computers, it should be of no practical disadvantage that the identified model's RFSs are given in form of lookup tables. Most available virtual simulation environments directly support this kind of definition of nonlinearities. Should such a definition be disadvantageous for some reason, one can easily fit a parametric function on the smoothed noisefree dataset using a simple parametric Least Squares Estimator (LSE). The choice of the proper analytical form of the RFSs at this point should not be a challenge anymore.

2.4 Summary of the nonparametric identification

The current section aims to give a structural summary of the nonparametric identification method that has been proposed in Chapter 2, which can be considered as the core of the present thesis. The summary is given as a list of the main steps that are necessary to carry out the proposed procedure, which is supported by an illustrated overview of the workflow in Figure 2.19.

In order to deploy the proposed synchronisation-based nonparametric identification method for nonlinear dynamic systems, carry out the following steps:

- Define the *N*DoF process model (i.e. the DEs that govern the dynamics of the system of interest). According to the subclass of mechanical systems defined in Section 1.3, which are given by (1.11), this means the definition of the inertia matrix $\mathbf{M}$ and the input matrix $\mathbf{B}$. (Chapter 4 offers a solution to identify these matrices in case they would not be known a priori.)

- Define the measurement setup (i.e. the measurement function $\mathbf{h}$) involving at least N measurement signals that satisfy (2.32) and (2.33).

- Carry out the measurements to generate time sequences of the input vector $\mathbf{u}$ and the measurement vector $\mathbf{y}$.

- Calculate the quantities $\mathbf{R}$, $\mathbf{x}_{a,0}$, $\mathbf{P}_{\mathbf{x}a,0}$ and $\mathbf{Q}$ using the formulas (2.48), (2.49), (2.50) and (2.51) respectively, which are necessary to run the Kalman Filter (KF).

- Determine the lower bound vector $\mathbf{s}^{\mathrm{L}}$ of the instantaneous effective stiffnesses according to (2.34) or set it to the general value of $10^{-2}\mathrm{E}(\mathbf{s})$ to parametrise the constraints of the KF defined in Section 2.2.5. The lower bound can be set to $-\infty$ in case of displacement measurements.

- Run the nonlinear KF algorithm as defined in Section 1.6.3 and 2.2.3 to obtain the time series of the estimates of the augmented state vector $\mathbf{x}_a$ and its covariance matrix $\mathbf{P}_{\mathbf{x}a}$. It is discussed in Section 2.2.3 under which conditions an Extended Kalman Filter (EKF) is deployable.

- Assign the global results of the KF to the N local Restoring Force Surfaces (RFS) of the system according to (2.56).

- Transform the local KF results of the jth RFS into probabilistic elastic and dissipative Restoring Force Curve (RFC) datasets, i.e. $\{\boldsymbol{\sigma}_E,\mathbf{P}_{\sigma E}\}_i$ and

$\{\boldsymbol{\sigma}_D,\mathbf{P}_{\sigma D}\}_i$, using (2.62), (2.64) and (2.65). Or generate the jth probabilistic coupled RFS dataset $\{\boldsymbol{\sigma},\mathbf{P}_{\sigma}\}_i$ using (2.69) and (2.70).

- To identify the jth RFS as an additive nonlinearity use the procedure described in Section 2.3.2 to generate the jth noisefree elastic and dissipative RFCs in form of the data vectors (lookup tables) $\boldsymbol{\sigma}^{\mathrm{S}}_{E,k}$ and $\boldsymbol{\sigma}^{\mathrm{S}}_{D,k}$ based on the datasets $\{\boldsymbol{\sigma}_E,\mathbf{P}_{\sigma E}\}_i$ and $\{\boldsymbol{\sigma}_D,\mathbf{P}_{\sigma D}\}_i$. Use the formula (2.84) to tune the smoothness of the characteristics. Use the uncertainty curves (2.85) and (2.86) to indicate the quality of the identified RFCs.

- To identify the jth RFS as a coupled nonlinearity use the procedure described in Section 2.3.3 to generate the jth noisefree RFS in form of the data vector (lookup table) $\boldsymbol{\sigma}^{\mathrm{S}}_{k}$ based on the dataset $\{\boldsymbol{\sigma},\mathbf{P}_{\sigma}\}_i$. Use $P^{\mathrm{S}}_{\sigma,k}$ defined as (2.103) to indicate the quality of the identified RFS.

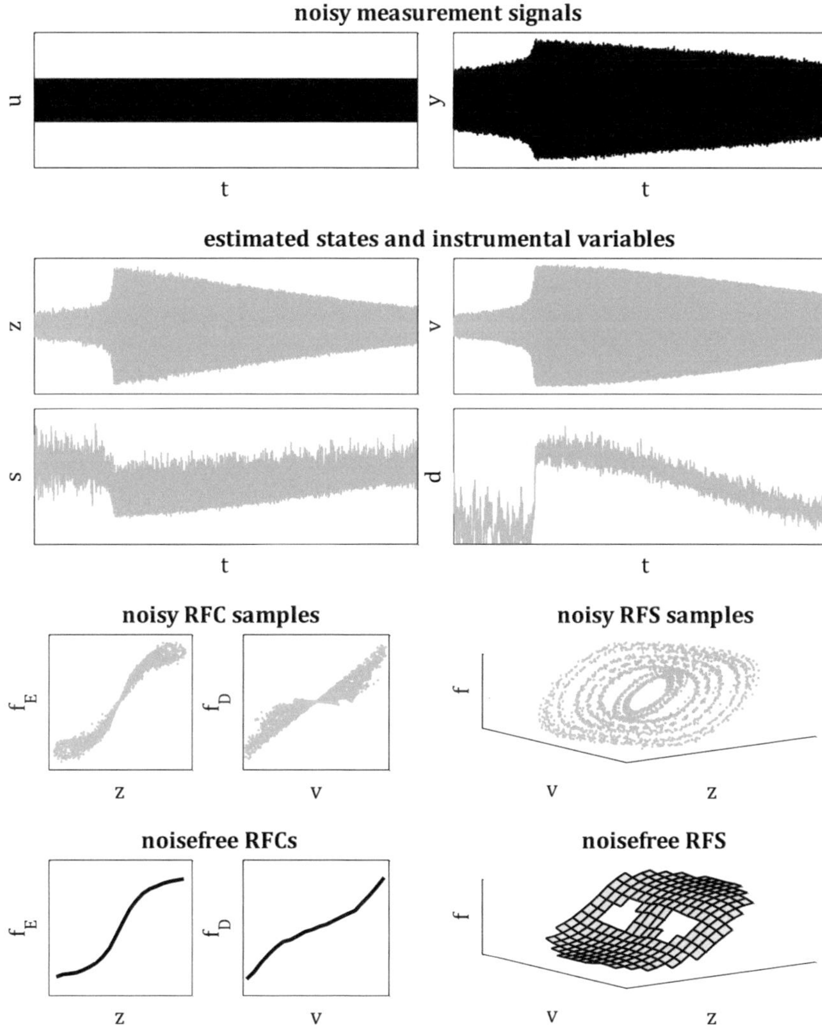

Figure 2.19: Workflow of the proposed synchronisation-based nonparametric identification method for nonlinear dynamic systems: Based on the measurement signals the Kalman Filter estimates the states and the instrumental variables (i.e. the instantaneous stiffness and damping). These are transformed into noisy samples of the RFCs and RFSs. Noisefree RFCs and RFSs are generated via optimal curve and surface fitting based on the noisy samples.

3 Virtual examples and the properties of the approach

The current chapter presents the nonparametric identification of several strongly nonlinear dynamic systems using the synchronisation based approach that has been proposed in Chapter 2. Each one of the chosen examples aims to point out some particular properties and aspects of the derived algorithm. Systems with 1DoF are investigated involving additive and coupled nonlinearities in Section 3.1 and 3.2 respectively, which are followed by a 3DoF mechanical structure in Section 3.3. The requirements of the technique on the measurement signals and its robustness against possible experiment issues is investigated in Section 3.4, followed by a brief analysis of the method's computational performance in Section 3.5.

The experiments are simulated virtually throughout this chapter. As long as it is not explicitly stated otherwise, the measurement signals are generated at a sampling rate of 10 kHz and are subsequently corrupted by 5 % RMS (q.v. *eRMS%* from Appendix B) uncorrelated Gaussian white noise, which represents typical realistic conditions for state-of-the-art test rigs in mechanical engineering. The excitation signal **u** is of kinematic type, i.e. it corresponds to the case $B_{ij} \in \{-1,0,1\}$ (q.v. Section 1.3).

The entire identification algorithm including the numerically generated measurement examples has been implemented in the MATLAB software environment (MathWorks, www.mathworks.com). The necessary analytical expressions have been derived using Maple (Maplesoft, www.maplesoft.com).

3.1 Additive restoring force surfaces

The present section investigates the identification of three strongly nonlinear 1DoF oscillators involving additive RFSs. The motion of the vibrating mass m is captured via noisy acceleration measurements. The true system equations are therefore of the form

$$\begin{bmatrix} \dot{z} \\ \dot{v} \end{bmatrix} = \begin{bmatrix} v \\ -m^{-1}\big(c_E(z) + c_D(v)\big) - u \end{bmatrix}, \tag{3.1}$$

$$y = -m^{-1}\big(c_E(z) + c_D(v)\big)\,. \tag{3.2}$$

Notice that the state variables (z and v) are chosen as the deformation and the rate of deformation of the system's RFS according to Section 1.3. The measurement equation (3.2) however corresponds to the acceleration of the body in the global coordinate system, which represents a real-life accelerometer. The a priori system model for the KF is defined as

$$\begin{bmatrix} \dot{z} \\ \dot{v} \\ \dot{s} \\ \dot{d} \end{bmatrix} = \begin{bmatrix} v \\ -m^{-1}p_c(sz + dv) - u \\ 0 \\ 0 \end{bmatrix}, \tag{3.3}$$

$$y = -m^{-1}p_c(sz + dv)\,, \tag{3.4}$$

which is of the form (2.31) and (2.33) with one modification: The total restoring force is scaled using the constant coefficient p_c (according to Section 2.2.4) in order to improve the conditioning of the state estimation error variance matrix $\mathbf{P}_{\mathbf{x}a}$, which improves the numerical stability of the KF. This modification influences the process function's Jacobian from equation (2.46) as well, as p_c appears in the derivatives of $\dot{v}$. The constant is chosen based on the expected average stiffness of the RFS. At the end of the identification procedure the smoothed RFCs are rescaled using the same coefficient. The kinematic excitation of the system, given by u, is carried out using a forward frequency sweep with constant amplitude of 5 ms^{-2}. The frequency range of the sweep is 100 Hz to 400 Hz in Subsection 3.1.1 and 3.1.3. It is slightly reduced in Subsection 3.1.2 to the range of 30 Hz to 250 Hz. The constant parameters are set to m=4 kg and p_c=5·10[6]. The tuning of the KF (i.e. the choice of $\hat{\mathbf{x}}_{a,0}^{+}$, $\widehat{\mathbf{P}}_{\mathbf{x}a,0}^{+}$, $\mathbf{Q}$ and $\mathbf{R}$) is carried out according to Section 2.2.4.

3.1.1 Comparison of additive and coupled identification

Since the current section includes the first identification example in the present thesis, its major aim is to prove the functionality of the nonparametric approach proposed in Chapter 2. Additionally the particular virtual example is used to compare the result quality of additive and coupled identification and to carry out a brief investigation of possible fine tuning of the KF. The true RFCs of the considered system are given by the following continuous symmetric nonlinear functions:

$$c_E(z) = 5 \cdot 10^5 \cdot z + 45 \cdot \tanh(4 \cdot 10^5 \cdot z) \tag{3.5}$$

$$c_D(v) = 50 \cdot v + 25 \cdot (100 \cdot v)^3 \tag{3.6}$$

The initial augmented state vector is set to $\hat{\mathbf{x}}_{a,0}^{+}$=[0,0,2 Nm^{-1},4·10^{-4} Nsm^{-1}]^{T}, where the scaled initial stiffness and damping values are chosen in order to represent a linearisation of the true system in the particular load case based on the measurement signal. A direct simulation of the a priori model with the initial values of the augmented states, from here on referred to as the "initial model", is compared to the true system in Figure 3.1 by means of the corresponding RFSs and measurement time series. The nonparametric identification of the system is carried out using the EKF. The lower bound constraint of the time-varying stiffness is chosen based on the initial model that has a damping ratio of $\mathcal{D}$=0.158, which leads to s^L=0.05 Nm^{-1} according to the formula (2.34). The derived approach allows the estimation of the system's nonlinear characteristics either as an additive RFS (as the sum of an eRFC and a dRFC) or as a coupled RFS. Accordingly, the identified system is compared to the true one based on the corresponding RFSs and measurement time series in Figure 3.2 and Figure 3.3 for the additive and the coupled identification case respectively. The frequency response of the identified system is simulated using the measured excitation signal *u*.

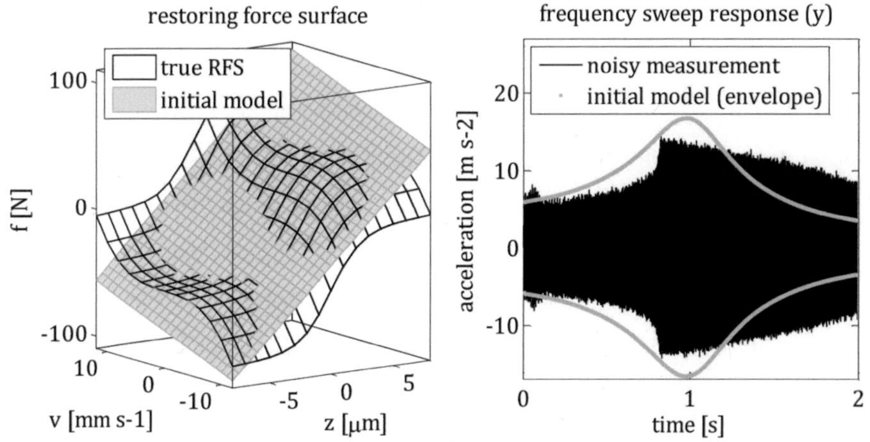

Figure 3.1: Comparison of the initial model and the true system based on the corresponding RFSs (left) and frequency sweep response time series (right).

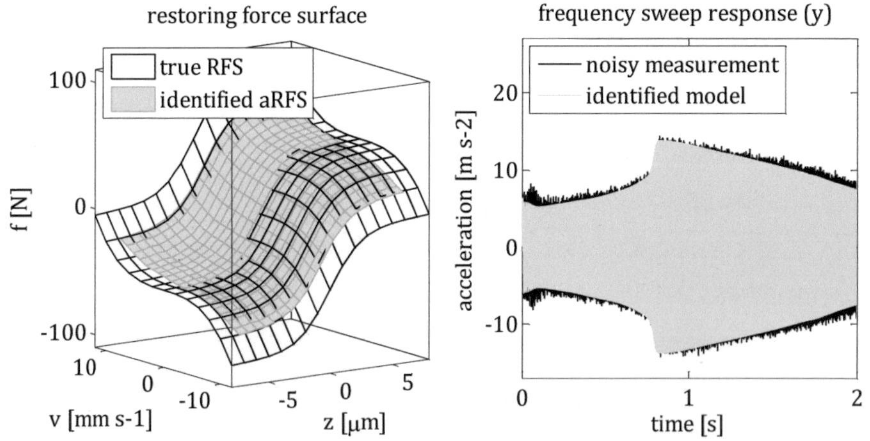

Figure 3.2: Comparison of the true and the identified system based on the corresponding RFSs (left) and frequency sweep response time series (right). The identification has been carried out assuming an additive nonlinearity (aRFS).

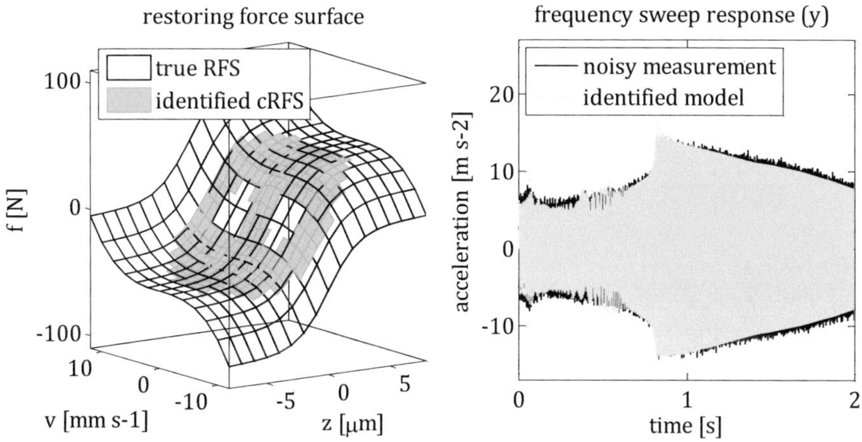

Figure 3.3: Comparison of the true and the identified system based on the corresponding RFSs (left) and frequency sweep response time series (right). The identification has been carried out assuming a coupled nonlinearity (cRFS).

By looking at the frequency sweep responses of the identified systems it can be observed that both approaches, i.e. the aRFS and the cRFS, deliver good identification quality, which is the main conclusion of the present section. Comparing the two results the following can be stated: On the one hand, only the cRFS managed to reconstruct the superharmonic resonance at the beginning of the frequency sweep. On the other hand, the aRFS results in much smoother system response. The eRFCs and dRFCs generated by the two approaches are depicted in Figure 3.4 in order to allow for a more precise comparison. Thereby two different techniques have been used to reduce the cRFS into an additive model. In the first case, referred to as "cRFS slice", simple sections of the cRFS are calculated at v=0 and z=0 to generate the eRFC and the dRFC respectively. In the second technique, referred to as "averaged cRFS slice", multiple sections at uniformly distributed v and z levels are calculated and subsequently debiased by the force values at z=0 and v=0 respectively. The average of these slices' force coordinates forms the final RFCs. It can be seen that the cRFS provides a significantly better estimate of the dissipative forces than the aRFS, however, the slice averaging technique is necessary to achieve sufficient smoothness of the characteristic.

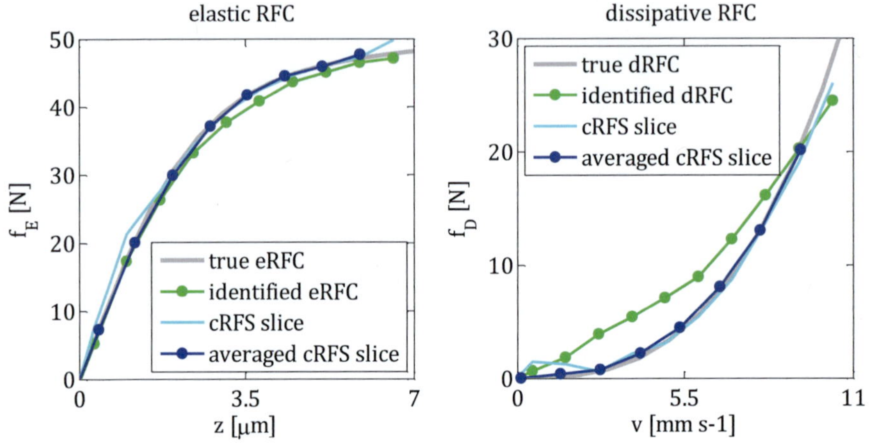

Figure 3.4: Comparison of additive and coupled identification based on the eRFCs (left) and dRFCs (right) of a 1DoF nonlinear system involving additive nonlinearity.

Notice that the identified cRFS in Figure 3.3 is only defined over the state space range, where the system's trajectory passed during the measurement. The empty segments ("holes") on the cRFS occur over those areas of the $\{z,v\}$ plane, where the lower bound for the number of samples per segment in the fitting algorithm has not been reached. The aRFS in Figure 3.2 is however valid over the whole rectangular range spanned by the maximum magnitudes of z and v that were reached during the frequency sweep. This validity range is a crucial property of the identified RFSs, which becomes clearer by considering the following example. Let us slightly modify the previously introduced virtual experiment setup. The eRFC given by (3.5) is kept unchanged, while the dRFC defined by (3.6) is replaced by a simple linear damping of 250 Nsm^{-1}. The excitation's amplitude is modified to 0.5 ms^{-2} and its frequency range is set to 200 Hz to 400 Hz. The nonparametric identification is carried out twice: first based on a forward sweep load case, and second using a backward sweep measurement. The corresponding reconstructed RFSs are referred to as "forward RFS" and "backward RFS" respectively. The response time series of the two identified systems are compared to the measured responses for both sweep directions in Figure 3.5. It can be observed that the forward response strongly differs from the backward response. This is a well known property of systems with nonlinear eRFCs, which has a significant influence on the validity range of the two identified RFSs. The comparison of the results reveals that the "forward RFS" is not completely valid for the backward sweep load case, because the latter covers a larger domain of the state space than the forward sweep. The "backward RFS" is however suitable to reconstruct both load cases with good accuracy.

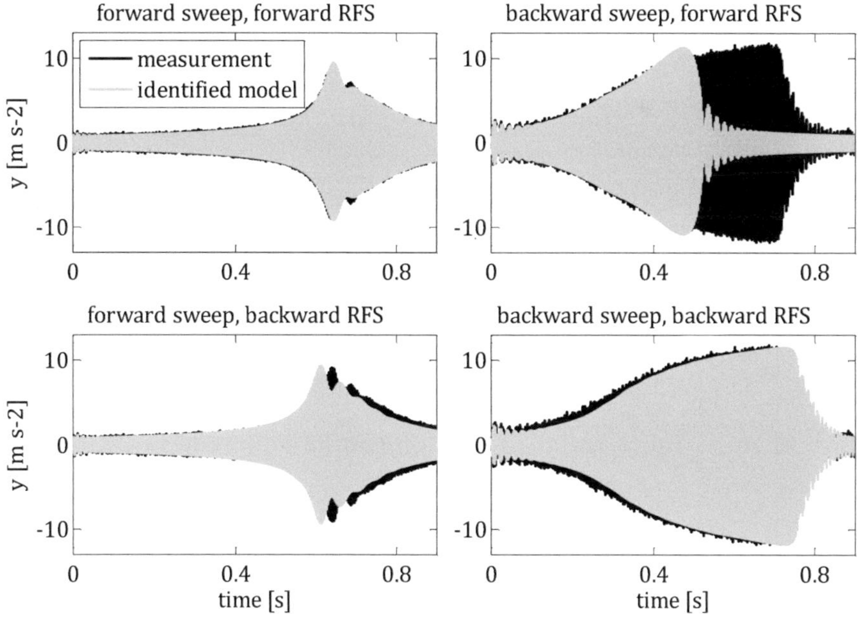

Figure 3.5: Measured forward (left) and backward (right) frequency sweep response time series of a 1DoF nonlinear system compared to the responses of two identified models. The first (top) has been identified during the forward sweep, while the second (bottom) is based on the backward sweep. Both sweeps has been carried out in the same frequency range of 200 Hz to 400 Hz.

Therefore, it can be concluded that the proper choice of the excitation plays an essential role in the resulting RFSs' applicability. This topic is further discussed on an example involving coupled nonlinearity in Section 3.2.

During the calculation of the presented results the process covariance exponents (q.v. Section 2.2.4) were set to q_s=3 and q_d=1. In order to investigate the influence of fine tuning on the EKF, a parameter study has been carried out, where the exponents have been varied in the range of $q_s \in [2,3.5]$ and $q_d \in [0,5.5]$ with a step size of 0.5. The effect of simultaneously increasing q_s and q_d has already been presented in Section 2.2.4 (q.v. Figure 2.7). In the current section the balance between the eRFC's and the dRFC's accuracy is of particular interest. Therefore, the results of the parameter study are depicted over the difference between the two exponents in Figure 3.6 in form of the *eRMS%* error of the RFCs.

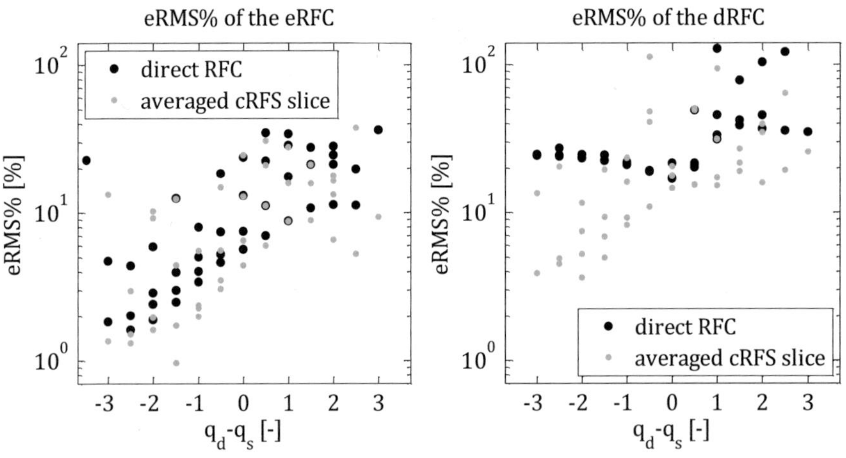

Figure 3.6: Results of the parameter study of q_s and q_d using the virtual frequency sweep measurement of a 1DoF system with additive nonlinearity. Additive identification („direct RFC") is compared to coupled identification („averaged cRFS slice") based on the *eRMS%* error (q.v. Appendix B) of the identified eRFC (left) and dRFC (right).

It can be concluded that setting q_d to smaller values than q_s is advantageous in general. It is however interesting that in case of additive identification (referred to as "direct RFC") the related quality increase is restricted to the eRFC. Whereas, in case of coupled identification the quality of the dRFC increases as well.

3.1.2 Identification of backlash

The aim of the current subsection is to demonstrate that the implemented constrained KF algorithm (q.v. Section 2.2.5) allows the identification of strongly nonlinear vibro-impact systems involving backlash based on acceleration measurements. The only difference of the considered virtual measurement setup from the one in Subsection 3.1.1 lies in its RFCs. They are defined in form of lookup tables and are depicted in Figure 3.8, which shows zero effective stiffness of the eRFC in a significant deformation range. The initial augmented state vector is set to $\hat{\mathbf{x}}_{a,0}^{+}=[0,0,1\ \mathrm{Nm}^{-1},2\cdot 10^{-4}\ \mathrm{Nsm}^{-1}]^{\mathrm{T}}$, which, similar to Subsection 3.1.1, represents an approximate linearisation of the true system in the particular load case. The initial model is compared to the true system in Figure 3.7. The identification is carried out using the EKF, where the lower bound constraint of s is set to $s^{\mathrm{L}}=0.0125\ \mathrm{Nm}^{-1}$ according to the formula (2.34) based on the initial model's damping ratio of $\mathcal{D}=0.112$. The identified RFCs are depicted in Figure 3.8, which includes the results of an additional simulation as well, where the displacement has been chosen as the measured quantity. The constraint s^{L} was set to $-\infty$ in this case, since according to Section 2.1.2 it is not needed for $y=z$.

It can be observed in Figure 3.8 that despite the observability problem of zero effective stiffness in case of $y\in\{v,\dot{v}\}$ (q.v. Section 2.1.2), the quality of the identified RFCs based on acceleration measurement is comparable to the one that is identified using the displacement signal. In fact the accuracy of the eRFC within the backlash is even higher in case of $y=\dot{v}$. To better point out the high quality of the identified characteristics, several time segments of the frequency sweep measurement are plotted in Figure 3.9. It compares the noisy measured signal to the reconstructed one, which was simulated using the RFCs that were identified from acceleration measurement.

In order to deeper understand how the constraint of the time-varying stiffness influences the estimation accuracy of the KF, the bias (offset) and the variance (noisiness) of the estimated time series of z and s (with respect to the true time series) are plotted in Figure 3.10 and Figure 3.11 for both acceleration (left) and displacement (right) measurements. It can be observed that the convergence of the displacement $z(t)$ is indeed lost if s^{L} is set to insufficiently low levels in case of $y=\dot{v}$. It is however apparent that the chosen value of s^{L} is ideal in the sense that it provides approximately the smallest possible bias of the instantaneous stiffness $s(t)$ for acceleration measurement. Furthermore, the figures explain the similar result quality of the RFCs in case of the two different measurement types, since the low bias of $s(t)$ for $y=z$ is compensated by its significantly higher variance compared to the case of $y=\dot{v}$. Although the curve fitting algorithm from Section 2.3.2 is designed to accommodate noisy samples, it has been shown in Section 2.2.1 that the highest achievable accuracy always remains connected to the variance of the samples.

Notice that negative values of s^{L} still influence the results, although the true values of s never become negative in the current example. The first reason for this is the significant variance of the estimated time-varying stiffness that can lead to negative values of the unconstrained estimates. The second reason lies in the probabilistic nature of the implemented constraining algorithm (q.v. Section 2.2.5), which results in the truncation of the augmented state vector's probability distribution even if the unconstrained mean value satisfies the constraint.

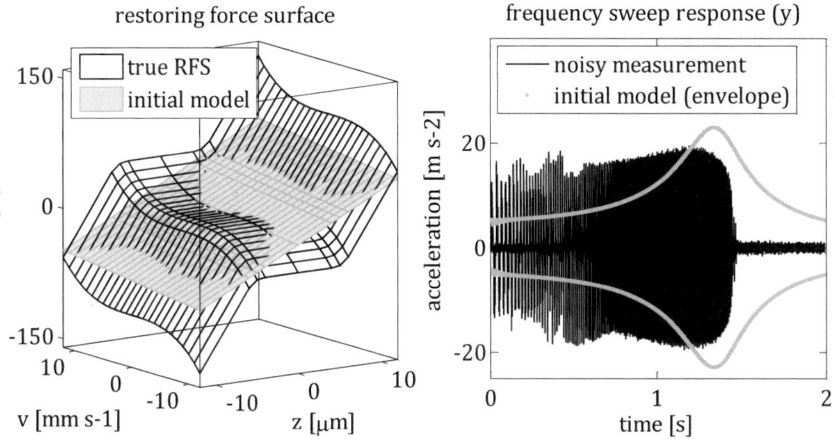

Figure 3.7: Comparison of the initial model and the true system based on the corresponding RFSs (left) and frequency sweep response time series (right).

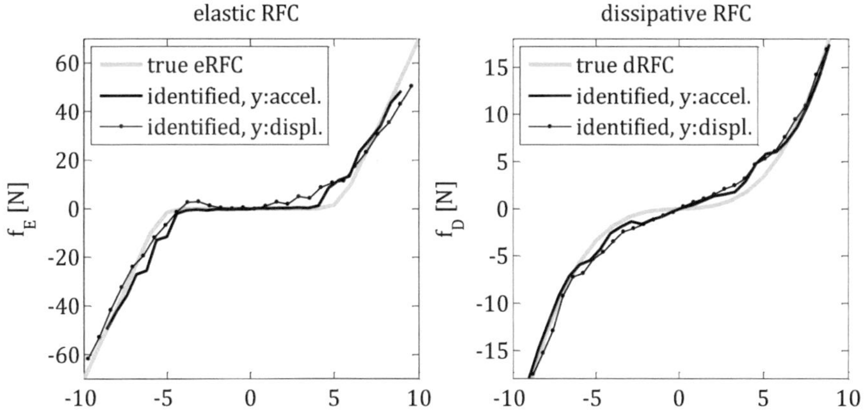

Figure 3.8: Comparison of the true and the identified eRFC (left) and dRFC (right) of a 1DoF nonlinear dynamic system involving backlash. The identification results for the two measurement types, displacement and acceleration, are compared.

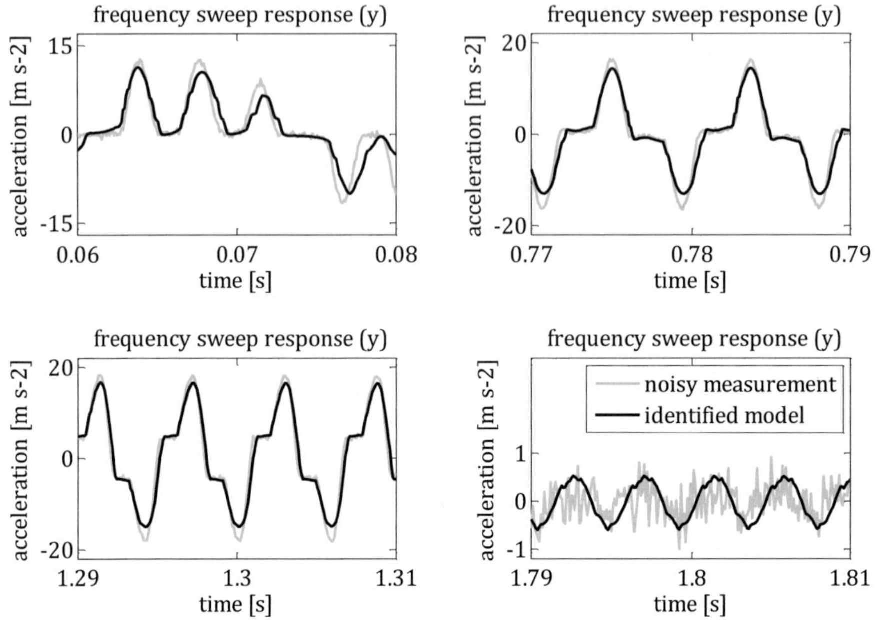

Figure 3.9: Comparison of the true and the identified system based on four different time segments of the corresponding frequency sweep response time series. The identification has been carried out using acceleration measurement.

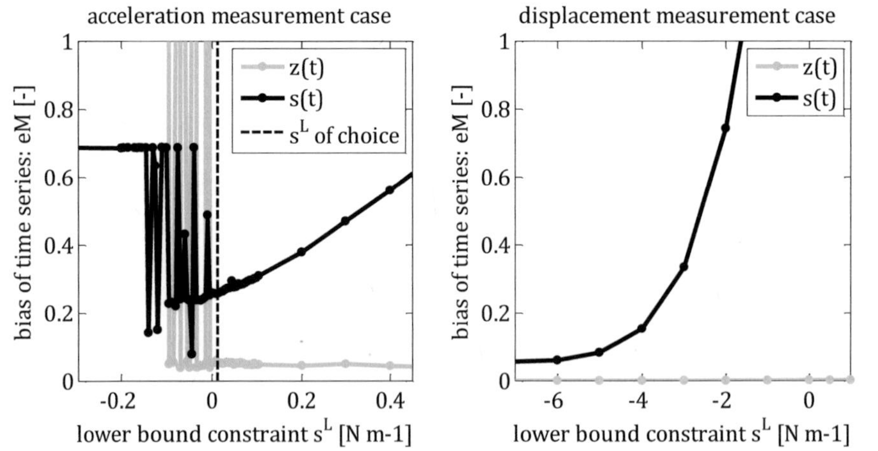

Figure 3.10: Bias (*eM*, q.v. Appendix B) of the time series of $z(t)$ and $s(t)$ with respect to their true values for different lower bound levels of the effective stiffness in case of acceleration (left) and displacement (right) measurement.

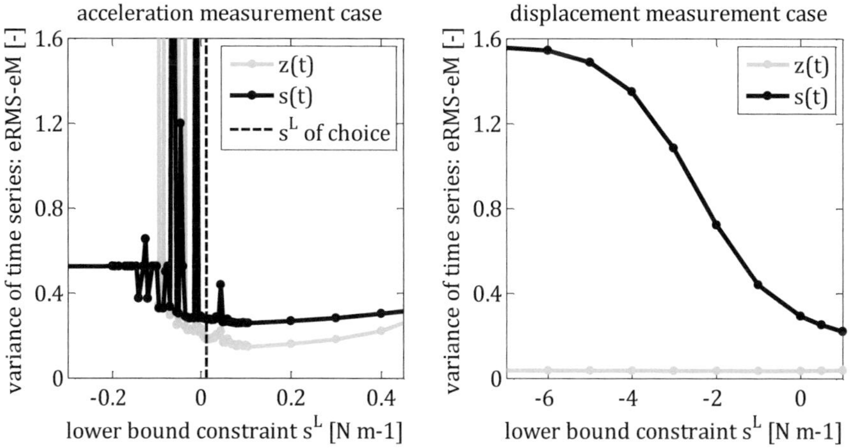

Figure 3.11: Variance (*eRMS–eM*, q.v. Appendix B) of the time series of $z(t)$ and $s(t)$ with respect to their true values for different lower bound levels of the effective stiffness in case of acceleration (left) and displacement (right) measurement.

3.1.3 Complex nonlinearity involving friction

The current subsection is dedicated to point out that, thanks to the weakly nonlinear general a priori model (2.31), the computationally efficient EKF algorithm can be deployed even in case of a complex asymmetric additive nonlinearity involving Coulomb friction. The virtual experimental setup is the same as in Subsections 3.1.1 and 3.1.2 except for the RFCs of the measured system. The tuning and constraining of the KF has also been done the same way as in the previous two subsections. Additive identification based on acceleration measurement has been carried out using the EKF as well as the SR-UKF, to which the corresponding identified RFCs are depicted in Figure 3.12 and Figure 3.13 respectively. The figures clearly prove that due to the weak bilinear nonlinearity of the process model, the SR-UKF does not achieve better accuracy than the EKF, therefore, there is no need for its significantly higher computational costs (q.v. Section 3.5). The current RFC plots utilise the uncertainty curves $\boldsymbol{\rho}^L$ and $\boldsymbol{\rho}^U$ (proposed in Section 2.3.2) that indicate the expectable local accuracy of the characteristics. Regardless of the type of the deployed KF it can be observed that the quality of the identified dRFC is much lower than in case of the two previous examples from Subsections 3.1.1 and 3.1.2. A possible explanation for this is that the dissipative forces are much lower compared to the elastic forces in the current case than they were in case of the other two systems. Notice that the absolute force deviation of the characteristics is of the same order of magnitude for the eRFC and the dRFC.

Further comparison of the two different KFs' accuracy is given by the frequency sweep response time series of the identified systems compared to the noisy measurement signal in Figure 3.14 for the EKF (left) and the SR-UKF (right). It can be seen that the EKF in fact provides slightly better result quality than the SR-UKF in this particular example.

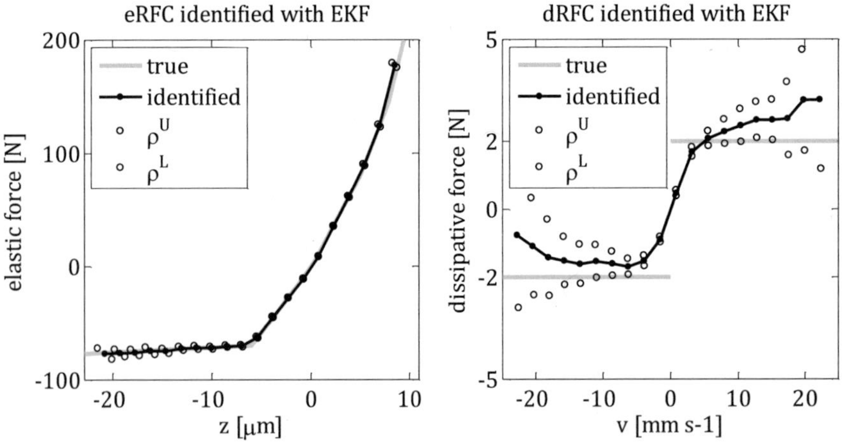

Figure 3.12: Comparison of the true and the identified eRFC (left) and dRFC (right) of a 1DoF dynamic system including the corresponding uncertainty curves $\boldsymbol{\rho}^L$ and $\boldsymbol{\rho}^U$. The identification was carried out based on acceleration measurement using the EKF.

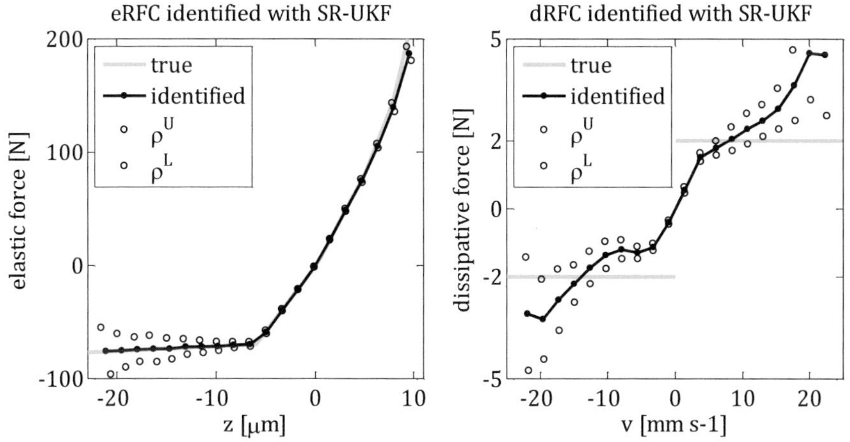

Figure 3.13: Comparison of the true and the identified eRFC (left) and dRFC (right) of a 1DoF dynamic system including the corresponding uncertainty curves $\boldsymbol{\rho}^{\mathrm{L}}$ and $\boldsymbol{\rho}^{\mathrm{U}}$. The identification was carried out based on acceleration measurement using the SR-UKF.

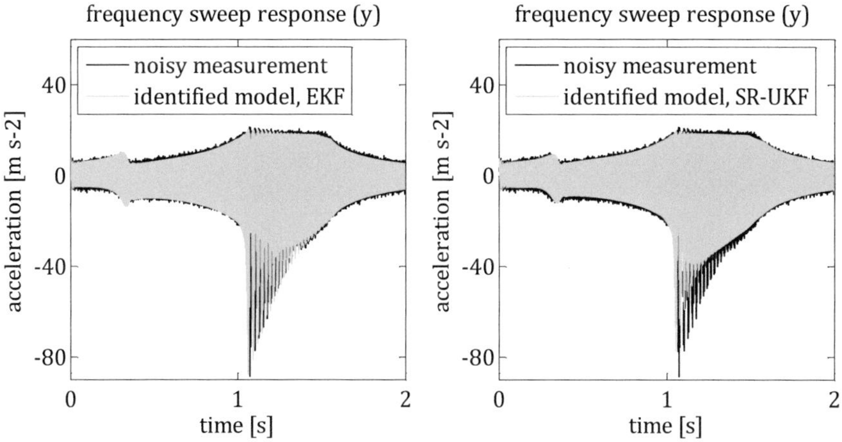

Figure 3.14: Comparison of the identified 1DoF system's frequency sweep response time series to the true measured response. The identification was carried out using the two KF types: EKF (left) and SR-UKF (right).

3.2 Coupled restoring force surfaces

Section 3.1 presented the nonparametric identification of three 1DoF dynamic systems involving strongly nonlinear additive RFSs. Although coupled identification has also been tested on some of these examples, the capability of the presented approach to identify coupled nonlinearities has not been presented so far. Therefore, the current section investigates the nonparametric identification of the Van der Pol oscillator, which is well known for its limit cycle that arises from its crucially coupled nonlinear nature (Strogatz, 1994). The reconstruction of such a system's characteristics from vibration measurements has already received some attention in the literature. (Sitz, Schwarz, Kurths, & Voss, 2002) can be mentioned as an example, where the parametric identification of a Van der Pol system is carried out using an augmented UKF.

Compared to Section 3.1 the only change in the virtual measurement setup is the replacement of the two RFCs by one cRFS in the system equations, which are therefore of the form

$$\begin{bmatrix} \dot{z} \\ \dot{v} \end{bmatrix} = \begin{bmatrix} v \\ -m^{-1}c(z,v) - u \end{bmatrix}, \tag{3.7}$$

$$y = -m^{-1}c(z,v)\,, \tag{3.8}$$

where the considered particular nonlinearity is given by the analytical function

$$c(z,v) = 5 \cdot 10^6 \cdot z + 1500 \cdot ((7 \cdot 10^4 \cdot z)^2 - 1) \cdot v\,. \tag{3.9}$$

The classical academic investigation considers the free system that exhibits self-excited oscillation, which converges to a stable limit cycle. This situation, i.e. $u(t)$=0, is investigated as the first load case. In contrast to the examples from Section 3.1 it is not possible to create an approximate linearisation of the system based on the available measurement. The EKF is therefore initialised with $\hat{\mathbf{x}}^{+}_{a,0}$=[0,0,1 Nm^{-1},10^{-4} Nsm^{-1}]^{T}, where the scaled stiffness and damping values represent typical initial guesses. The lower bound constraint of s is simply set to s^L=0.01 Nm^{-1}, which proved to be a good rule of thumb during the present study. The identified cRFS and the true state space trajectory of the system during the measurement are depicted in Figure 3.15.

After the identification the adapted model has been used to reconstruct the measured behaviour. The initial state was thereby set to the small nonzero values of $z_0=10^{-6}$ m and $v_0=10^{-3}$ ms^{-1} in order to introduce a disturbance that aims to initiate the expected free oscillations. Although the accuracy of the reconstructed cRFS is very good, the free oscillations of the identified system deviate significantly from the measured behaviour, which can be observed in Figure 3.17 (left). The reason for this is the general property of nonparametric cRFSs that they are only valid over the state space domain, where the system spent sufficient time during the measurement. The proper choice of the experiment load case, as discussed e.g. in (Link, Boeswald, Laborde, Weiland, & Calvi, 2011), is therefore crucial for the nonparametric identification of such systems.

The forced vibrations of the Van der Pol oscillator are investigated in (Mohamed, Karim, & Belghith, 2013), where the external excitation is shown to have significant influence on the system's behaviour. Accordingly, a harmonic excitation with an amplitude of 15 ms^{-2} and a constant frequency of 100 Hz has been introduced as a modified load case in the current study in order to reach a much wider range of the $\{z,v\}$ plane. The true state space trajectory of the forced system during the measurement and the identified cRFS are depicted in Figure 3.16. The system that was identified based on the modified load case is now able to reconstruct the true trajectories of the measured system's free oscillations with good accuracy, which is depicted in Figure 3.17 (right).

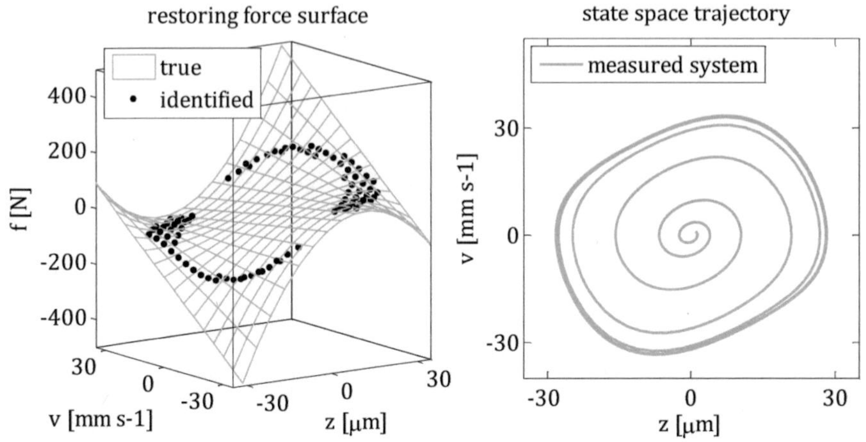

Figure 3.15: True and identified cRFS (left) of the Van der Pol oscillator and the state space trajectory (right) of the measured virtual system during a load case without external excitation, i.e. $u(t)=0$.

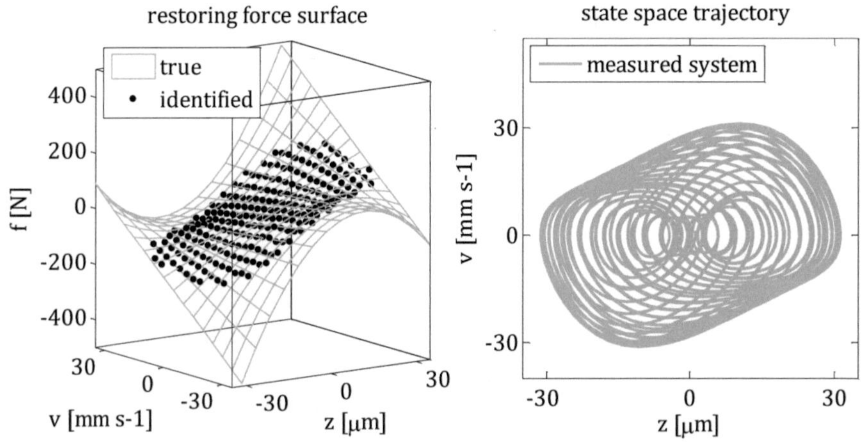

Figure 3.16: True and identified cRFS (left) of the Van der Pol oscillator and the state space trajectory (right) of the measured virtual system during a load case with an external excitation of $u(t)=15\sin(2\pi 100t)$ ms^{-2}.

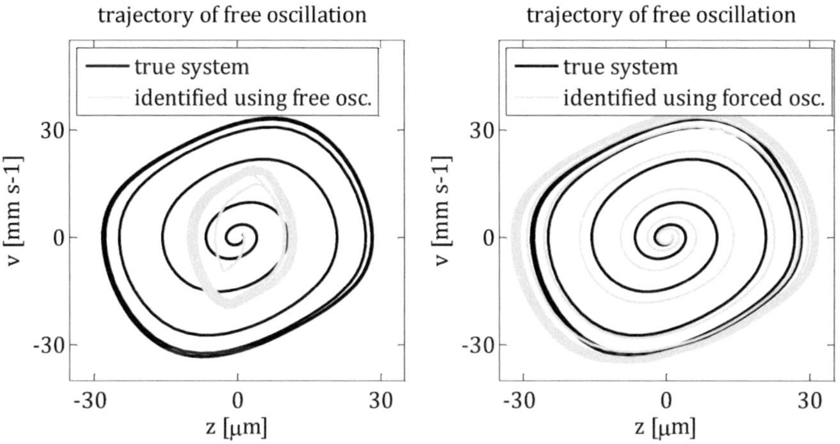

Figure 3.17: Comparison of the identified Van der Pol system's free oscillation to the true limit cycle for the two different identification load cases: free oscillation (left) and forced vibration (right).

3.3 Multiple DoF systems

According to the problem formulation from Section 1.3 the present study considers not only 1DoF systems but addresses the identification of multiple DoF structures, particularly of the form (1.11), as well. To investigate this capability of the proposed approach, the identification of the three nonlinear RFSs of a 3DoF mechanical system based on noisy virtual measurements is presented in the current section. Let us begin with the introduction of the virtual experiment setup.

3.3.1 A 3DoF plate

Several real-life engineering structure's dynamics can be approximately described by a rigid body with one translational and two tilting DoFs linked to its surroundings via elastic coupling elements. Such a model can be deployed to investigate passenger car body vibrations, the isolation of machine foundations and the oscillation of clutch components (Tikhomolov, 2015), just to mention a few of the possible applications. The virtual measurement setup of such a system is illustrated in Figure 3.18, on which the identification example is based in the following subsection. The three DoFs of the depicted rigid plate are translation in ζ direction and rotation around the coordinate axes ξ and η, denoted by φ_ξ and φ_η, assuming small tilting angles. The body is mounted to the ground via three nonlinear bipolar spring elements, whose behaviour is described by their RFSs given by the functions $c_j(z_j,v_j)$. The connection points of the springs to the plate are given by the coordinate triplets $\{\xi_1,\eta_1,0\}$, $\{\xi_2,\eta_2,0\}$ and $\{\xi_3,\eta_3,0\}$. The inertia properties of the plate in the $(\zeta,\varphi_\xi,\varphi_\eta)$ coordinate system are given by its diagonal inertia matrix of the form

$$\mathbf{M}_{(\zeta,\varphi_\xi,\varphi_\eta)} = \begin{bmatrix} m & 0 & 0 \\ \cdot & J_\xi & 0 \\ \cdot & \cdot & J_\eta \end{bmatrix}, \tag{3.10}$$

where m is the mass of the body and J_ξ and J_η denote the mass moment of inertia around the axes ξ and η respectively. The structure is excited via the vibration of the ground given by the acceleration signal u. According to Section 1.3 the deformation and the rate of deformation of the spring elements, i.e. z_j and v_j, are chosen as the state space coordinates of the a priori system model for the identification. This results in an a priori process model of the form (2.31) with 12 state variables, i.e. n_a=12. The inverse inertia matrix $\mathbf{M}^{-1}$ and the input matrix $\mathbf{B}$ are given as

$$\mathbf{M}^{-1}_{(z_1,z_2,z_3)} = \frac{1}{m}\begin{bmatrix} 1 & 1 & 1 \\ \cdot & 1 & 1 \\ \cdot & \cdot & 1 \end{bmatrix} + \frac{1}{J_\xi}\begin{bmatrix} \eta_1^2 & \eta_1\eta_2 & \eta_1\eta_3 \\ \cdot & \eta_2^2 & \eta_2\eta_3 \\ \cdot & \cdot & \eta_3^2 \end{bmatrix} + \frac{1}{J_\eta}\begin{bmatrix} \xi_1^2 & \xi_1\xi_2 & \xi_1\xi_3 \\ \cdot & \xi_2^2 & \xi_2\xi_3 \\ \cdot & \cdot & \xi_3^2 \end{bmatrix}, \tag{3.11}$$

$$\mathbf{B}_{(z_1,z_2,z_3)} = \begin{bmatrix} -1 \\ -1 \\ -1 \end{bmatrix}. \tag{3.12}$$

The measurement vector **y** consists of the acceleration signals of the connection points of the spring elements in ζ direction, denoted y_j, which leads to the measurement equation

$$\mathbf{y} = \begin{bmatrix} y_1 \\ y_2 \\ y_3 \end{bmatrix} = \begin{bmatrix} \dot{v}_1 + u \\ \dot{v}_2 + u \\ \dot{v}_3 + u \end{bmatrix}. \tag{3.13}$$

This set of signals satisfies the necessary measurement conditions defined by (2.32) and (2.33) as long as there are no coinciding sensor points. Notice that for the sake of clarity equation (3.13) is given as a function of state derivatives, which however need to be replaced by the corresponding rows of the process equation (2.31) prior to the implementation in the KF algorithm.

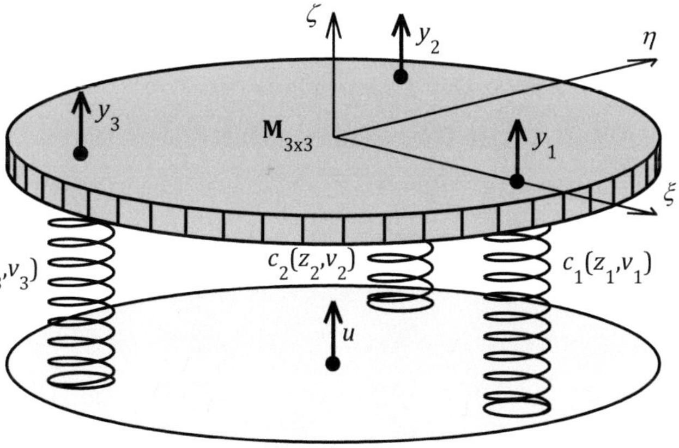

Figure 3.18: Virtual experiment setup of a rigid plate on elastic foundation with one translational DoF in ζ direction and two tilting DoFs around the ξ and η coordinate axes. The three bipolar spring elements represent general nonlinear RFSs between the ground and the body.

3.3.2 Identification of mounting preload

The current subsection presents the nonparametric identification of a 3DoF plate using the virtual experiment setup introduced in Subsection 3.3.1. To give this example a practical meaning, let us consider the realistic situation, where the three spring elements represent the mounting of the body to the vibrating ground. Let us suppose that the plate is a rigid body model of a machine that is not allowed to lift off from the ground under some expected vibration conditions. In this case one can use high level excitation of the system to identify the effective preload in each of the three mounting segments. To do so, a backward frequency sweep input has been applied to the virtual structure with constant amplitude of 7 ms^{-2} in a frequency range of 500 Hz to 100 Hz. The properties of the system were set to:

$$\mathbf{M}^{-1}_{(z_1,z_2,z_3)} = \begin{bmatrix} 0.408 & 0.046 & 0.046 \\ \cdot & 0.408 & 0.046 \\ \cdot & \cdot & 0.408 \end{bmatrix} [\mathrm{kg}^{-1}] \,, \tag{3.14}$$

$$\begin{bmatrix} \xi_1 & \xi_2 & \xi_3 \\ \eta_1 & \eta_2 & \eta_3 \end{bmatrix} = 0.13 \cdot \begin{bmatrix} \cos(0°) & \cos(120°) & \cos(240°) \\ \sin(0°) & \sin(120°) & \sin(240°) \end{bmatrix} [\mathrm{m}] \,. \tag{3.15}$$

Instead of the approximate linearization of the system in the particular load case (q.v. Section 3.1) the EKF was simply initialised with the typical guess (3.16). The scaling coefficient of the RFSs (q.v. Section 3.1) was set to p_c=10^7. Since the formula (2.34) is not convenient for multiple DoF systems, the lower bound constraint of **s** was defined with general entries as (3.17), which proved to be a good rule of thumb throughout the current thesis.

$$\hat{\mathbf{x}}^{+}_{a,0} := \begin{bmatrix} [\,0 & 0 & 0\,]^{\mathrm{T}} & [\mathrm{m}] \\ [\,0 & 0 & 0\,]^{\mathrm{T}} & [\mathrm{ms}^{-1}] \\ [\,1 & 1 & 1\,]^{\mathrm{T}} & [\mathrm{Nm}^{-1}] \\ [10^{-4} & 10^{-4} & 10^{-4}]^{\mathrm{T}} & [\mathrm{Nsm}^{-1}] \end{bmatrix} \tag{3.16}$$

$$\mathbf{s}^{\mathrm{L}} = [10^{-2} \quad 10^{-2} \quad 10^{-2}] \ [\mathrm{Nm}^{-1}] \,. \tag{3.17}$$

The additive identification of the three nonlinear RFSs (including single-sided preload) has been carried out using the measurement time series **y** corrupted by 5 % RMS (q.v. *eRMS%* from Appendix B) zero mean white noise. The results are depicted in Figure 3.19. Apparently there is a very good agreement between the true characteristics and the identified RFSs. Hence, it can be stated that the estimated eRFCs are suitable for the determination of the effective local preload in the three mounting segments. Furthermore, it can be observed in Figure 3.20 that the identified model is able to reconstruct the measured system's complex nonlinear response with good accuracy.

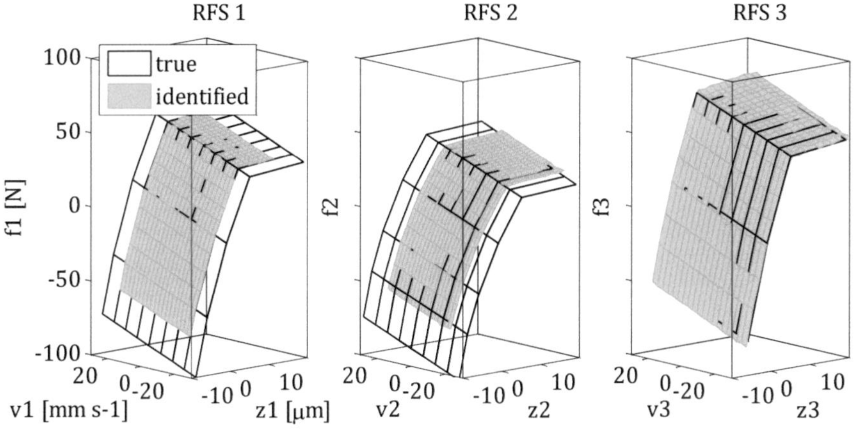

Figure 3.19: Comparison of the true and the identified RFSs of the 3DoF mechanical structure. The identification was carried out in its additive form using the EKF.

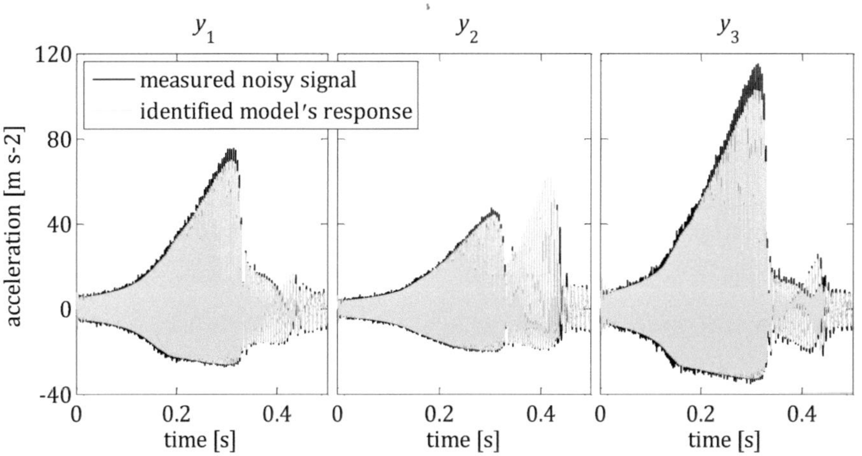

Figure 3.20: Comparison of the measured behaviour of the 3DoF mechanical system to the reconstructed frequency sweep response using the identified model.

3.4 Requirements on the measurement

In order for an identification technique to be suitable for real-life application, its stability under realistic measurement conditions is of high importance. The current section therefore investigates the robustness of the proposed approach against typical experiment issues such as high noise level (q.v. Subsection 3.4.1), poor sampling rate (q.v. Subsection 3.4.2) and model uncertainties (q.v. Subsection 3.4.3).

3.4.1 Noise level

It has been discussed in Section 2.2.1 that noise is an inevitable component of every measured signal, which in most cases can not be neglected. It can be described by its bias vector and its discrete-time covariance matrix **R**. Significantly biased measurements can not be handled by the presented technique. They can drastically decrease result quality or even cause observer divergence. Fortunately in most situations the bias can either be avoided or detected and compensated. It is however important to mention that the correction of the signal drift of frequency sweep measurements of nonlinear systems has to be treated carefully. It is shown e.g. in (Tikhomolov, 2015) that such signals exhibit a natural drift of their time average in case of strongly asymmetric elastic characteristics, which should not be altered by the signal correction. In contrast to bias, the a priori correction of the noise covariance, which is practically speaking the actual noisiness of the signal, always corrupts the higher harmonic components of nonlinear oscillations. That is the reason why the KF has been chosen as the synchronisation algorithm in the present study, since it can accommodate noisy signals in a proper probabilistic manner.

To show the resilience of the technique to noise, the identification of the virtual measurement example from Section 3.1.1 has been carried out at different RMS levels of the zero mean Gaussian white noise that is added to the signal y. The noisy measurements and the eRFCs, reconstructed using the EKF in combination with additive identification, are depicted in Figure 3.21. The comparison of the results for different noise levels highlights the advantage of the probabilistic manner of the presented approach. It can be observed that even for an extremely high noise level of 50 % (q.v. *eRMS%* from Appendix B), the technique still yields useful results.

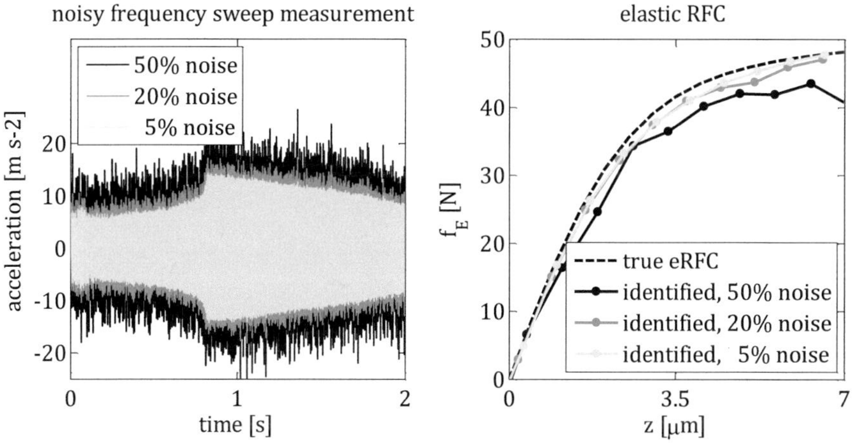

Figure 3.21: Comparison of the reconstructed eRFCs (right) of a 1DoF nonlinear system. The additive identification has been carried out based on frequency sweep measurements (left) at different *eRMS%* (q.v. Appendix B) noise levels using the EKF.

In order to gain deeper insight into the algorithm's behaviour in the presence of noise, a comparison of different subtypes of the approach is presented in Figure 3.22 based on the estimation error of the identified RFS of the 1DoF system from Section 3.1.1. The subfigure on the left shows the results as a function of the noise level applied to the measurement signal *y*. It can be seen that in case of additive identification, referred to as aRFS, the noise has a relatively weak influence on the accuracy. However, the quality of the RFS based on coupled identification, referred to as cRFS, is much stronger affected. It can be observed that the cRFS technique delivers better results for moderate noise levels, which has already been investigated in Section 3.1.1. On the other hand, the cRFS solution becomes less accurate than the aRFS for high noise levels. This can be explained by the slower statistical convergence of the cRFS, which is discussed later in detail in Subsection 3.4.2. Furthermore, it can be observed that the SR-UKF exhibits approximately the same dependence on the noise of y as the EKF.

The effect of the input signal's noisiness is depicted in the right subfigure of Figure 3.22. It is to be seen that the EKF has a much weaker resistance to this kind of disturbance compared to the noise of the measurement signal. The reason for this lies in the main concept of the proposed tuning of the EKF. Recall from Section 2.2.4 that, in order to achieve fast convergence of the instrumental variables' estimates ($\mathbf{s}$ and $\mathbf{d}$) over time, all the entries of $\mathbf{Q}$ that are not related to their derivatives has been set to zero, q.v. formula (2.51). Hence, the non-augmented part of the a priori process model, i.e. the describing ODEs of the mechanical system, are assumed to be correct, which tends to ignore any error of the input signal. It is interesting however that the SR-UKF is able to compensate the noisiness of u nearly as good as it accounts for the noise in y. This is probably a result of the fact that in this technique (in contrast to the EKF) the input signal has a direct effect on the calculated covariance matrices through the sigma points of the state vector. This shows that, though the SR-UKF is computationally more expensive than the EKF, it is a much more sophisticated algorithm. Fortunately the noise content of the excitation (input) signal is in most cases relatively low, since it is a characteristic that is typically directly controlled during the measurement. Furthermore, in case of frequency sweep input the precise filtering of the noise from the signal is possible, since the expected form of the noisefree time series is well known.

The above investigation considered zero mean Gaussian white noise. This is a justified assumption in most real-life situations. More general formulations of the KF that can account for noise processes with certain correlation over time, so called "coloured noise", also exist (Simon, 2006). This topic is however outside the scope of the present thesis.

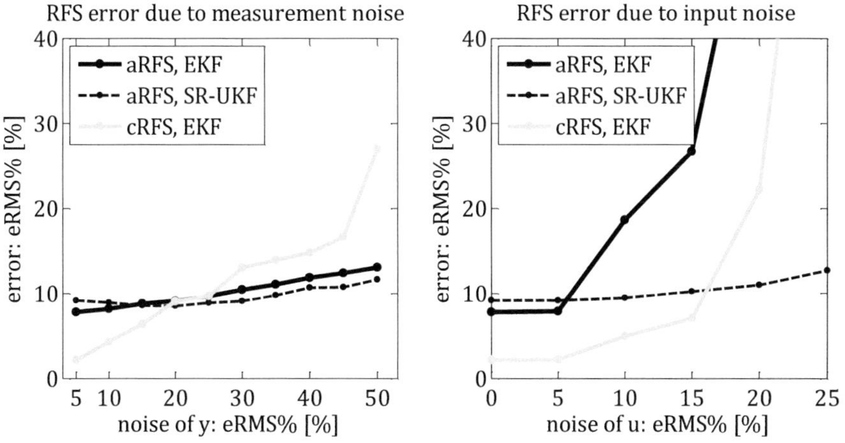

Figure 3.22: Comparison of different subtypes of the proposed nonparametric identification algorithm based on the estimation error of the identified RFS of the 1DoF system from Section 3.1.1 as a function of the noise level applied to the measurement signal y (left) and to the input signal u (right).

Before the investigation of further measurement related aspects, let us briefly discuss the possible application of the presented nonparametric approach to time-delay systems, since (as described in the following) this topic is closely related to the noisiness of the input signal. The complex dynamics of such systems received notable attention in the past decades (Stépán, Szalai, & Hogan, 2005), (Insperger & Stépán, 2002), (Stépán, 1989). They can be described via delay differential equations (DDEs), which in case of a 1DoF nonlinear oscillator with a single time delay, denoted by τ_D, are of the form

$$\begin{bmatrix} \dot{z}(t) \\ \dot{v}(t) \end{bmatrix} = \begin{bmatrix} v(t) \\ -m^{-1}c(z(t), v(t)) + bz(t - \tau_D) \end{bmatrix}. \tag{3.18}$$

The literature already offers some identification techniques that can estimate both the time delay and further nonlinearities of such systems (Voss & Kurths, 1997). The method that is explored in the present study is not able to identify τ_D. However, the question arises, whether it can be deployed to identify the RFS, i.e. $c(z,v)$, of such a system supposing that the time delay is known a priori. If we consider $z(t-\tau_D)$ as the input signal u, then the structure of the system is covered by the problem formulation from Section 1.3. Let us assume that a good initial guess of the time series of z is available for the time period $t\epsilon[-\tau_D,0]$ in order to initialise the KF. In this case the only remaining critical aspect is that instead of the exact signal of u only its noisy estimate is provided, which is generated by the KF itself. Notice that this leads to the same problem that has been discussed previously in the current section, i.e. the noisiness of u, as long as the estimates of z remain approximately unbiased. Though this topic is not further investigated in the current thesis, it can be stated based on the results depicted in Figure 3.22 that the application of the presented nonparametric identification approach to time-delay systems seems possible. Finally, it should be mentioned that a special extension of the KF, the so called Fixed-Lag Smoother (Simon, 2006), could be implemented to enhance the stability of the algorithm. This technique allows the recursive refinement of the estimate of $z(t_i-\tau_D)$ based on the measurements in the time interval $t\epsilon[t_i-\tau_D,t_i)$.

3.4.2 Sampling

Measurement signals of today are always digitally sampled. The three most important consequences of this fact are the following:

- The value of each sample has a finite numerical precision.
- The signal has a finite sampling rate over time.
- There are only a finite number of measured samples available.

The first can typically be accounted for via proper calibration of the sensors during the experiment. It can therefore be treated as an effect without potential risks regarding the identification procedure. The second, i.e. the finite sampling rate, however plays an important role in the accuracy and stability of synchronisation based methods especially in the current case of nonparametric identification. Due to the time-varying stiffness and damping coefficients, the a priori system model diverges from the true system between the correction steps of the KF, i.e. between measurement sampling times (q.v. Section 2.2.3 and 2.2.4). The longer the time step between two measurement samples, the less accurate the algorithm becomes. To investigate this effect, the 1DoF oscillator from Section 3.1 was considered with an additive nonlinearity of the form

$$c(z,v) = 2 \cdot 10^{6} \cdot z + 2 \cdot 10^{17} \cdot z^{3} + 1250 \cdot v\,. \tag{3.19}$$

The virtual system has been excited using a harmonic input signal with constant amplitude of 5 ms^{-2} and constant frequency of 250 Hz. The acceleration of the oscillating mass has been measured using different sampling rates and subsequently corrupted by 5 % (q.v. *eRMS%* from Appendix B) zero mean white noise. The duration of the measurement variants has been adapted to maintain 10^4 time steps in order to separate this investigation from the topic of the number of samples. The estimation error of the identified RFSs (using additive identification) is depicted in the left subfigure of Figure 3.23 comparing the accuracy of the EKF and the SR-UKF. The results are illustrated as a function of the sampling frequency divided by the main oscillation frequency (i.e. 250 Hz), which is referred to as the "normed sampling rate". It can be concluded that decreasing sampling frequency leads to lower identification quality. Drastic increase of error can be observed if the normed sampling rate becomes significantly lower than 20. Furthermore, the results show that there is no clear difference between the accuracy of the EKF compared to the SR-UKF. However, the SR-UKF remains convergent for such low sampling rates, where the EKF already exhibits synchronisation divergence.

The third consequence of digital sampling, i.e. the finite number of samples, is also essential in case of nonparametric identification. Although it does not affect the estimation quality of the KF, it does however influence the accuracy of the subsequential RFS fitting. To explore this influence, the virtual measurement of the previous investigation has been carried out with different durations at fixed sampling rate of 10 kHz in order to vary the number of samples. The excitation has been changed to a forward frequency sweep from 100 Hz to 400 Hz to enable coupled identification as well (q.v. Section 3.2). Special synthetic KF results have been generated to allow the separate analysis of the RFS fitting procedure. This means that the synchronisation bias of the KF has been artificially eliminated from the time series of the augmented state vector. Furthermore, the estimates of the deformation and the rate of deformation (z and v) have been completely freed from noise to enable the comparison of the proposed approach to parametric Least Squares (LS) fitting (Amiri-Simkooei & Jazaeri, 2012), which can not take the noisiness of the RFS's independent variables into account. The fitting quality of the derived additive and coupled nonparametric approaches is compared to the accuracy of polynomial parametric LS fitting in the right subfigure of Figure 3.23 as a function of the number of incorporated samples. The resolution of the estimated RFS's lookup table was set to 20 segments for both the z and the v coordinate.

The results clearly point out an important difficulty of nonparametric fitting, which can be summarised as follows. In contrast to global regression (based on an assumed parametric function) each node of the generated nonparametric lookup table is mainly defined by a reduced local group of observed samples. This increases the influence of the slow statistical convergence of such a "natural Monte Carlo sampling" (q.v. Section 2.2.1). In case of the additive fitting of RFCs the number of the mainly influencing points for one local fitted node reduces in linear relation with the increase of the resulting lookup table's resolution. This relation becomes quadratic in case of coupled surface fitting of RFSs, which further amplifies the undesirable effect of slow statistical convergence. Of course the gap between the different fitting techniques' accuracy reduces proportionally to the noise level of the samples. The resolution of the lookup table is also an important influencing factor, which has a problem specific optimum regarding the bias and the variance of the results. As a simple guideline it can be concluded, that the necessary number of samples to achieve good fitting quality is about 10^3 for additive and about $2 \cdot 10^4$ for coupled non-parametric fitting for a resolution of 10 to 20 segments per dimension. This means a measurement duration of at least 2 to 3 seconds for typical measurements of high frequency oscillations for coupled fitting. Finally, it is important to notice that this difficulty has nothing to do with the estimation of the RFS sample point clouds using the KF. These sample sets enable an easy characterisation of the RFSs for parametric fitting in case of insufficient number of samples.

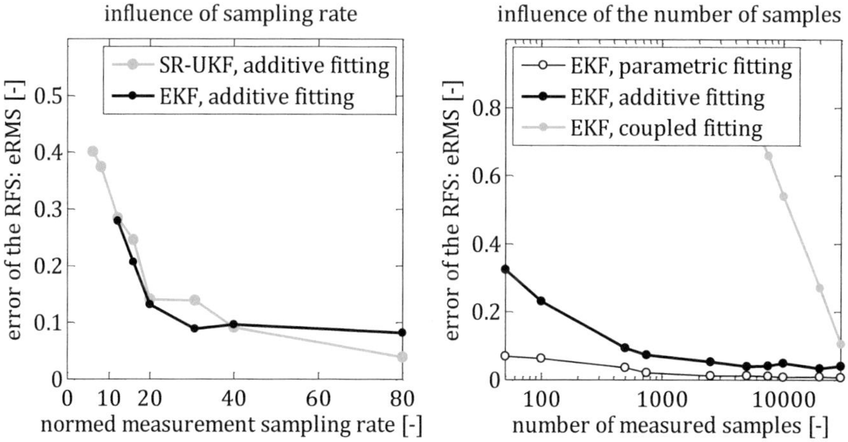

Figure 3.23: Investigation of the influence of measurement sampling rate (left) and the number of measured samples (right) on the estimation error of the identified RFS based on the noisy (5 % RMS) virtual measurement of a 1DoF nonlinear oscillator. The resolution of the reconstructed RFS's lookup table was set to 20 segments per dimension.

3.4.3 Model uncertainties

It has been discussed in Section 1.6.3 that synchronisation algorithms (and therefore the KF as well) are sensitive to modelling uncertainties. This means that the deviation of model parameters (that have not been introduced as augmented state variables) from their true values can reduce the accuracy and the stability of such techniques. The current subsection briefly investigates this effect in the particular case of nonparametric identification using the virtual 3DoF example from Section 3.3.2. To do so, let us introduce the following notation for the entries of the inverse inertia matrix $\mathbf{M}^{-1}$ given in the coordinate system of the deformation of the RFSs, i.e. z_j:

$$\mathbf{M}_{(z_1,z_2,z_3)}^{-1} := \begin{bmatrix} \mu_{11} & \mu_{12} & \mu_{13} \\ \cdot & \mu_{22} & \mu_{23} \\ \cdot & \cdot & \mu_{33} \end{bmatrix}. \tag{3.20}$$

Different kinds of errors of the assumed values of μ_{ij} have been introduced in the numerical example to analyse their influence on result quality. The corresponding identified eRFCs of the system are depicted in Figure 3.24 and Figure 3.25, which show a clear decrease of estimation accuracy due to the artificial modelling errors.

There are several further possibilities of how the a priori model can deviate from reality. The unexpected noisiness of the input signal (q.v. Section 3.4.1) is one of these possibilities. An issue that can be even more critical than parameter uncertainty is the incorrect choice of the number of DoFs. Such errors can lead to completely unrealistic identification results or even to synchronisation divergence. This fact increases the importance of developing computationally efficient identification algorithms, which was one of the primary aims of the present thesis (q.v. Section 1.1). Short computation time is essential, because it allows the implementation of the algorithm inside a higher level iteration procedure that successively eliminates modelling errors. The investigation of this topic in general is outside the scope of the current study. Nevertheless, the particular problem of uncertain model parameters, which is probably the most typical modelling error, is briefly investigated in Chapter 4, which offers a possible solution for the identification of the inverse inertia matrix $\mathbf{M}^{-1}$.

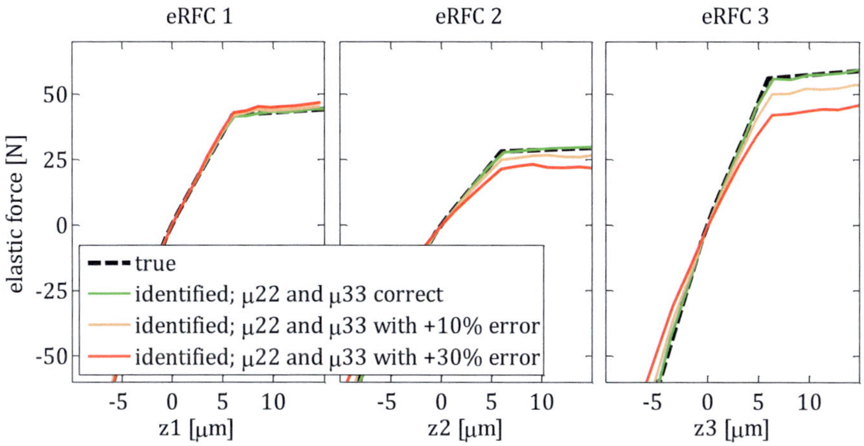

Figure 3.24: Influence of the incorrect values of the inverse inertia matrix's diagonal entries on the estimation accuracy of the 3DoF virtual system's three eRFCs from Section 3.3.2.

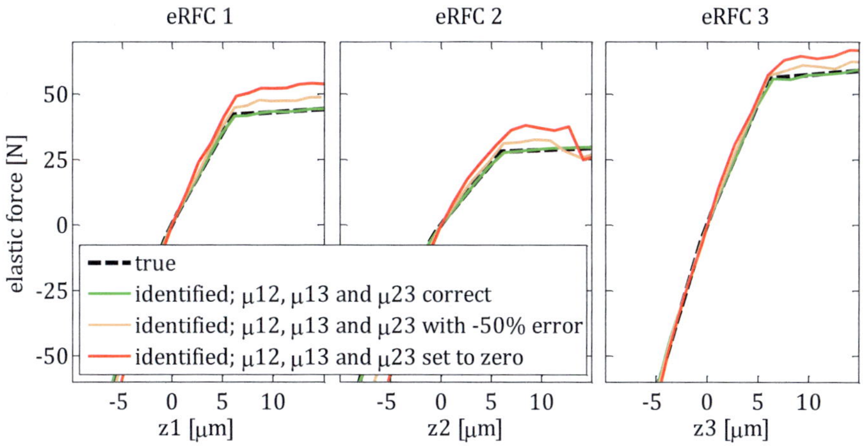

Figure 3.25: Influence of the incorrect values of the inverse inertia matrix's off-diagonal entries on the estimation accuracy of the 3DoF virtual system's three eRFCs from Section 3.3.2.

3.5 Computational performance

A major motivation to the proposed nonparametric identification technique was to achieve a computationally efficient algorithm (q.v. Section 1.1). As discussed in Section 1.4, there are available parametric identification methods that in general can identify nonlinear systems of arbitrary complexity. This however can lead to enormous computation times in case of missing a priori knowledge on the nonlinearities of the system of interest. Hence, numerical simplicity due to simple general models, such as the one proposed in Section 2.1.2, represents the main advantage of nonparametric methods. Therefore, the current section presents a brief analysis of this topic, in order to point out the presented algorithm's computational performance.

There are several possibilities of accelerating a numerical procedure that are not directly related to the algorithm itself. The most trivial one is the deployment of faster computers. But there are also more sophisticated solutions such as parallel computing or the generation of symbolically optimised codes. In order to separate the current performance evaluation from these topics and to enable easier comparison to other methods, the "normed CPU time", denoted by T_{NC}, is introduced as a measure of the computation's duration. It is defined as the ratio between the specific procedures CPU time and the duration of a single simulation run of the initial a priori model under the given experiment conditions. E.g. in case of an optimisation process, every call of the objective function corresponds to a normed CPU time of 1. This quantity is used in Figure 3.26 to illustrate the computational efficiency of separate steps of the proposed nonparametric approach based on the virtual examples from Section 3.1 and 3.3. The indicated values are thereby independent from both the sampling rate and the duration of the available measurement signals, since the CPU time of all steps and of the initial model run are both proportional to the number of processed samples.

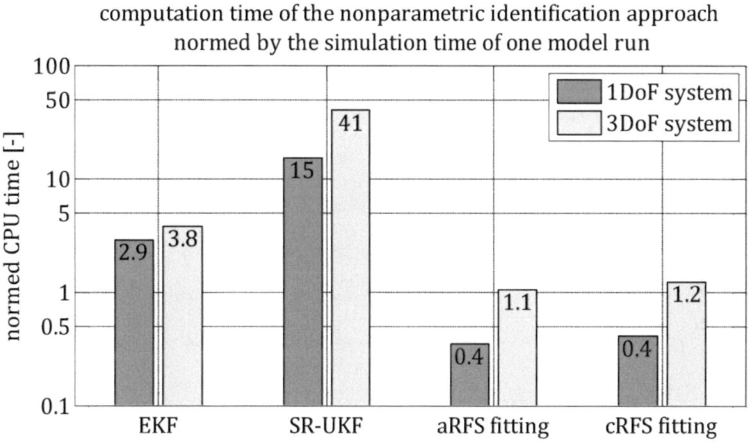

Figure 3.26: Computational efficiency of the proposed nonparametric technique's separate steps based on the identification examples from Section 3.1 and 3.3.

Notice that not all four depicted steps are necessary to carry out a particular identification. One application of the algorithm consists of either the EKF or the SR-UKF followed by either aRFS fitting or cRFS fitting. The T_{NC} level of the SR-UKF exhibits approximately linear dependency on the number of DoFs and shows significantly higher values than the EKF. This can be explained by the necessary number of $2n_a+1$ sigma points in the algorithm (q.v. Section 1.6). It has been extensively discussed in Chapter 2 and 3 that an important advantage of the proposed nonparametric method is that it allows the deployment of the EKF under typical experiment conditions despite the presence of strong nonlinearities in the system of interest. On the other hand, in order to cover a wider range of possible measurement circumstances, it has also been investigated in detail under which conditions the application of the SR-UKF is advised. Taking a look at the CPU times of the RFS fitting reveals that the T_{NC} of the fitting step is linearly proportional to the number of RFSs and is less than 0.5 per RFS for both approaches.

Without presenting any particular results using parametric identification techniques, the illustrated T_{NC} statistics can be used to carry out a rough comparison of the proposed approach to such methods. Let us first consider the KF based dual state and parameter estimation (q.v. Section 1.4.1). In this method every describing parameter of the a priori characterised RFS is introduced as an additional augmented state variable. Accordingly, as soon as the nonlinear function requires more than two parameters, the corresponding KF algorithm will necessarily be slower then the presented one, which uses only s and d for one RFS. The computation gets even more complex if general functions (e.g. polynomial series) have to be deployed due to the lack of a priori characterisation of the RFSs. Furthermore, the introduced functions of the augmented states are as nonlinear as the RFSs that they describe, which automatically leads to the necessity of computationally expensive KF methods such as the SR-UKF. This can be observed e.g. in (Wu & Smyth, 2007). The comparison to optimisation based methods (q.v. Section 1.4.1) is also straightforward. In these techniques the objective function of the optimisation is based on the simulation results of the virtual model, whose parameters are to be identified. Therefore, each function call during the optimisation increases the T_{NC} level by 1. (Vyasarayani, Uchida, Carvalho, & McPhee, 2012) and (Gunnarsson, 2014) can be mentioned as successful applications of the Homotopy optimisation method. In the first a necessary number of 50 and 250 iterations have been reported to reach convergent identification results for dynamic systems with two and three parameters respectively. In the latter 1400 iterations of the optimiser were needed to identify the four coefficients of the describing third-order polynomial of a dynamic system's nonlinear spring. The required number of function calls per iteration depends on the specific optimisation algorithm. However, even if we assume a T_{NC} increase of 1 per iteration, such approaches still show to be computationally more expensive than nonparametric identification. A more detailed comparison of optimisation based techniques to observer based recursive methods can be found in (Voss, Timmer, & Kurths, 2004).

4 Parametric identification of the a priori model

The previous two chapters presented the synchronisation based nonparametric identification of a particular subclass of dynamic systems given by the process equation (1.11). Throughout these chapters all constant parameters of the system, i.e. the entries of the inertia matrix **M** and the input matrix **B**, has been assumed to be known a priori. In real-life implementation such constants always exhibit a certain level of uncertainty. According to Section 1.6.3 and 3.4.3, errors in the assumptions of these parameters can lead to drastic decrease of the identified RFS's accuracy. Hence, it is of practical importance to provide proper estimates of these constants. One way to achieve this is the coupling of the proposed nonparametric approach with parametric identification techniques. This possibility is briefly investigated in Section 4.1 for the particular 3DoF virtual experiment from Section 3.3.2.

In situations, where the virtual model is thought of as an image of the true system's behaviour in the particular experiment load case, the determination of the structure's initial state, i.e. $\mathbf{z}_0$ and $\mathbf{v}_0$, is a part of the identification problem. This gains more importance e.g. for the measurements of free oscillations, where the reproduction of the measured behaviour using the adapted model is impossible without the proper initial state vector. This topic has been neglected so far in the previous chapters, since the KF (in contrast to optimisation based methods) automatically compensates the initial deviations of the state space coordinates. In Section 4.2 a modification of the KF is applied to the virtual example from Section 3.3.2, which allows the identification of the system's initial state.

The presented solutions throughout this chapter point out an important advantage of using the KF algorithm for synchronisation. Due to its potential and popularity, several extensions and special formulations have been developed throughout the last more than fifty years (q.v. Section 1.6.3). The choice of the KF makes all these features available, which leads to a more general and flexible identification procedure.

4.1 Estimation of inertia properties

Let us consider dynamic processes given by (1.11) and suppose that only rough guesses of **M** and **B** are available prior to the measurement. According to the problem formulation from Section 1.3, the input matrix **B** either consists of specific entries of $\mathbf{M}^{-1}$ (if **u** is a force signal) or $B_{ij}\in\{-1,0,1\}$ holds for its elements (if **u** is an acceleration signal). Hence, the identification of **M** instantly determines **B** as well. Therefore, it is sufficient to concentrate on the estimation of the inertia matrix. Notice that the proposed nonparametric system model (2.31) can not be fitted to the measured signals by simply adapting **M** using a parametric approach. Only those methods can be considered as possible extensions that can accommodate the complete nonparametric KF algorithm (derived in Chapter 2). This is crucial in order to ensure the simultaneous synchronisation of **s** and **d**, while the parametric algorithm successively determines the entries of **M**.

One parametric technique that fulfils this criterion is the Homotopy optimisation approach (q.v. Section 1.4.1). Though successful applications in the literature show its good reliability (Gunnarsson, 2014), (Vyasarayani, Uchida, Carvalho, & McPhee, 2012), (Carlsson & Nordheim, 2011), (Sun & Yang, 2010), it is a computationally rather expensive method due to its optimisation based algorithm (q.v. Section 3.5). Therefore, another suitable technique, the dual state and parameter estimation (q.v. Section 1.4.1), has been chosen in the present study, which has a higher potential for computational efficiency. It is based on the augmentation of the a priori model's state vector by the vector of unknown (uncertain) parameters, denoted by **p**. Notice that this is the same solution, which has been applied to enable the identification of RFSs in Chapter 2 with the difference that **p** does not consist of instrumental variables but physical model parameters, and that its true values are not expected to vary over time. Accordingly, the general a priori system model, proposed in Section 2.1.2, is now modified to

$$\begin{bmatrix}\dot{\mathbf{z}}\\ \dot{\mathbf{v}}\\ \dot{\mathbf{s}}\\ \dot{\mathbf{d}}\\ \dot{\mathbf{p}}\end{bmatrix} = \begin{bmatrix}\mathbf{v}\\ -\mathbf{M}^{-1}(\mathbf{p})[\mathrm{diag}(\mathbf{s})\mathbf{z} + \mathrm{diag}(\mathbf{d})\mathbf{v}] + \mathbf{B}(\mathbf{p})\mathbf{u}\\ \mathbf{0}\\ \mathbf{0}\\ \mathbf{0}\end{bmatrix}, \tag{4.1}$$

$$\mathbf{w}_{N\times 1} := [w_1, \dots, w_N]^{\mathrm{T}} \ni w_j \in \{z_j, v_j, \dot{v}_j\}, \tag{4.2}$$

$$\mathbf{y}_{N\times 1} = \mathbf{h}(\mathbf{w}, \mathbf{u}) \ni \mathrm{rank}\left(\frac{\partial \mathbf{h}}{\partial \mathbf{w}}\right) = N. \tag{4.3}$$

Hence, the structure of the KF's estimates of the state vector's probability distribution properties also changes from (2.41) to

$$\mathbf{X}_a = \begin{bmatrix}\mathbf{Z}\\ \mathbf{V}\\ \mathbf{S}\\ \mathbf{D}\\ \mathbf{P}\end{bmatrix} \sim \left(\mathbf{x}_a = \begin{bmatrix}\mathbf{z}\\ \mathbf{v}\\ \mathbf{s}\\ \mathbf{d}\\ \mathbf{p}\end{bmatrix}, \mathbf{P}_{\mathbf{X}a} = \begin{bmatrix}\mathbf{P}_\mathbf{Z} & \mathbf{P}_\mathbf{ZV} & \mathbf{P}_\mathbf{ZS} & \mathbf{P}_\mathbf{ZD} & \mathbf{P}_\mathbf{ZP}\\ \cdot & \mathbf{P}_\mathbf{V} & \mathbf{P}_\mathbf{VS} & \mathbf{P}_\mathbf{VD} & \mathbf{P}_\mathbf{VP}\\ \cdot & \cdot & \mathbf{P}_\mathbf{S} & \mathbf{P}_\mathbf{SD} & \mathbf{P}_\mathbf{SP}\\ \cdot & \cdot & \cdot & \mathbf{P}_\mathbf{D} & \mathbf{P}_\mathbf{DP}\\ \cdot & \cdot & \cdot & \cdot & \mathbf{P}_\mathbf{P}\end{bmatrix}\right), \tag{4.4}$$

where **P** denotes the probability distribution of **p**. Furthermore, the initialisation and the tuning of the KF also have to be extended. These modifications follow the same logic that has been discussed in Section 2.2.4, which leads to the formulas:

$$\hat{\mathbf{x}}_{a,0}^{+} := [\ \mathrm{E}(\mathbf{Z}_0)^{\mathrm{T}} \quad \mathrm{E}(\mathbf{V}_0)^{\mathrm{T}} \quad \mathrm{E}(\mathbf{S})^{\mathrm{T}} \quad \mathrm{E}(\mathbf{D})^{\mathrm{T}} \quad \mathrm{E}(\mathbf{P})^{\mathrm{T}}\ ]^{\mathrm{T}}, \tag{4.5}$$

$$\widehat{\mathbf{P}}_{\mathbf{X}a,0}^{+} := \mathrm{diag}([\ \mathrm{E}(\mathbf{Z}^2)^{\mathrm{T}} \quad \mathrm{E}(\mathbf{V}^2)^{\mathrm{T}} \quad \mathrm{E}(\mathbf{S}^2)^{\mathrm{T}} \quad \mathrm{E}(\mathbf{D}^2)^{\mathrm{T}} \quad \mathrm{E}(\mathbf{P}^2)^{\mathrm{T}}\ ]), \tag{4.6}$$

$$\mathbf{Q} := \Delta t \cdot \mathrm{diag}([\,\mathbf{0}_{1\times N} \quad \mathbf{0}_{1\times N} \quad 10^{q_s}\mathrm{E}(\mathbf{S}^2)^{\mathrm{T}} \quad 10^{q_d}\mathrm{E}(\mathbf{D}^2)^{\mathrm{T}} \quad 10^{q_p}\mathrm{E}(\mathbf{P}^2)^{\mathrm{T}}\,]), \tag{4.7}$$

where q_p is an additional covariance exponent that allows the separate convergence tuning of the estimates of **p**. Considering the particular case from Section 3.3.2 the entries of the inverse inertia matrix $\mathbf{M}^{-1}$ are defined as the elements of **p**, which using the notation from Section 3.4.3 leads to

$$\mathbf{p} := [\,\mu_{12} \quad \mu_{13} \quad \mu_{22} \quad \mu_{23} \quad \mu_{33}\,]^{\mathrm{T}}\,. \tag{4.8}$$

The first diagonal entry of $\mathbf{M}^{-1}$, i.e. μ_{11}, has not been defined as state variable. The reason for this is that the excitation is of kinematical type (acceleration) in the given virtual experiment setup. Under such conditions the inertia matrix of a mechanical system is only identifiable up to an unknown coefficient. This means that an endless number of systems exist that can reproduce the given measurement signals, which are therefore not distinguishable based on the particular experiment setup. In the current case μ_{11} has been set to its true value and has been kept constant, in order to enable the convergence of the other parameters to their correct values. In real-life one can achieve convergent results using the above model, but these will not necessarily be consistent. However, the missing coefficient of $\mathbf{M}^{-1}$ can be determined e.g. by measuring the weight of the rigid body. Further discussions on similar distinguishability issues can be found in (Dresig & Fidlin, 2014), (Bessa, Hackbarth, Kreuzer, & Radisch, 2014), and (Hoshiya & Saito, 1984).

It is important to mention that the defined form of the a priori system model from Section 3.3 is optimal for the handling of possible model uncertainties. The following aspects have been taken into account during its design, which can be considered as general guidelines for the modelling of multiple DoF systems for identification: Though the system has only 3 DOFs, the complete description of the experiment setup includes a relatively large number of parameters. The rigid body's three inertia properties (m, J_ξ, J_η) and one $\{\xi,\eta\}$ coordinate pair for the position of each sensor and each spring element adds up to 15 describing constants, all of which can be affected by some errors. Due to the choice of the spring elements' deformations and rate of deformations as state space coordinates and due to the positioning of the sensors at the spring elements' connection points, all possible parameter uncertainties, i.e. all 15 constants, have been "compressed" into the 6 entries of **M**. First, this leads to a drastic decrease of model complexity. Second, the fact that this "compression" is possible shows that the 15 original constants are not distinguishable and would therefore lead to convergence issues in the synchronisation algorithm. Of course, placing the sensors at the spring positions might not always be possible in real-life. Or, to go further, there might be cases, when there are more spring elements than DoFs, which automatically leads to indistinguishability. A simple and pragmatic solution for these situations is the assumption of N spring elements at the N sensor positions (q.v. Section 5.2.2). This leads to a virtual system that is at least able to reconstruct the measured behaviour. The problem of transforming the determined RFSs into properties of the real spring elements can be treated separately after the identification. The second important aspect that influenced the design of the model was the aim to avoid strong nonlinearities in the KF to allow the deployment of computationally simpler algorithms, i.e. the EKF instead of the SR-UKF (q.v. Section 2.2.2). The choice of the entries of $\mathbf{M}^{-1}$ instead of the elements of **M** as augmented states avoids the introduction of strong hyperbolic nonlinearities (e.g. $z_j s_j m^{-1}$) in the model and leads to less critical trilinear expressions (e.g. $z_j s_j \mu_{ij}$) instead.

Unfortunately it turned out that both the EKF and the SR-UKF are not able to achieve sufficient convergence of **p** in a single KF run due to the nonlinearity of the model and the large number of 17 state variables (z_j, v_j, s_j, d_j, μ_{ij}). To overcome this difficulty, the global iteration of the KF has been implemented in the present study. This is a well known technique with successful implementation examples in the literature (Voss, Timmer, & Kurths, 2004), (Hoshiya & Saito, 1984). The main idea is to iterate the complete KF algorithm by setting the last estimate of the state vector from the previous KF run as the initial state of the next run. This procedure only makes sense in case of dual state and parameter estimation and should not be confused with the locally iterated EKF (Simon, 2006), which iterates the model's Jacobians inside the recursion loop. Using the notations from expression (4.4), the precise definition of the implemented iteration is given by the following simple algorithm:

$$\hat{\mathbf{x}}^{+}_{a,0,k+1} = \begin{bmatrix} \mathbf{z}_{0,1} \\ \mathbf{v}_{0,1} \\ \mathbf{s}_{,\mathrm{end},k} \\ \mathbf{d}_{,\mathrm{end},k} \\ \mathbf{p}_{,\mathrm{end},k} \end{bmatrix}, \tag{4.9}$$

$$\hat{\mathbf{P}}^{+}_{\mathbf{X}a,0,k+1} = \begin{bmatrix} \mathbf{P}_{\mathbf{Z},0,1} & \mathbf{P}_{\mathbf{ZV},0,1} & \mathbf{0} & \mathbf{0} & \mathbf{0} \\ \cdot & \mathbf{P}_{\mathbf{V},0,1} & \mathbf{0} & \mathbf{0} & \mathbf{0} \\ \cdot & \cdot & \mathbf{P}_{\mathbf{S},\mathrm{end},k} & \mathbf{P}_{\mathbf{SD},\mathrm{end},k} & \mathbf{P}_{\mathbf{SP},\mathrm{end},k} \\ \cdot & \cdot & \cdot & \mathbf{P}_{\mathbf{D},\mathrm{end},k} & \mathbf{P}_{\mathbf{DP},\mathrm{end},k} \\ \cdot & \cdot & \cdot & \cdot & \mathbf{P}_{\mathbf{P},\mathrm{end},k} \end{bmatrix}, \tag{4.10}$$

where k denotes the index of the global iteration, and the recursion index of the last measurement time step is indicated by "end". Notice that the initial guesses of **z** and **v** remain unchanged throughout the global iteration, since their values are not getting refined by the KF.

The final step in the preparation of the parameter estimation algorithm is the choice of the KF's tuning coefficients. An extensive parameter analysis has been carried out using the given measurement setup, which yielded the set of tuning exponents $[q_s,q_d,q_p]=[-4,-6,-4]$ as the optimal choice. These optimised settings can be treated as a rough general guideline, since the introduced KF tuning strategy (4.7) is based on a normed formula. Recall from Figure 2.7 that such settings lead to a high bias of the estimates of **s** and **d**, since they are fastly varying. This however represents no difficulty, because the only aim of this identification step is the estimation of the parameter vector **p**. The identification of the RFSs is carried out in a second step using suitable tuning strategy according to Section 2.2.4.

Since $\mathbf{M}^{-1}$ is a constant linear property, its elements do not depend on the oscillation amplitude of the system. Therefore, it is reasonable to introduce an additional load case with low excitation level and compare the corresponding estimation results with the ones that are achieved using the original load case from Section 3.3.2. It can be expected that the low excitation level leads to better convergence of the parameter vector **p** due to the reduced oscillation of **s** and **d** over time. The response of the 3DoF plate to frequency sweep excitation is depicted in Figure 4.1 for the two load cases.

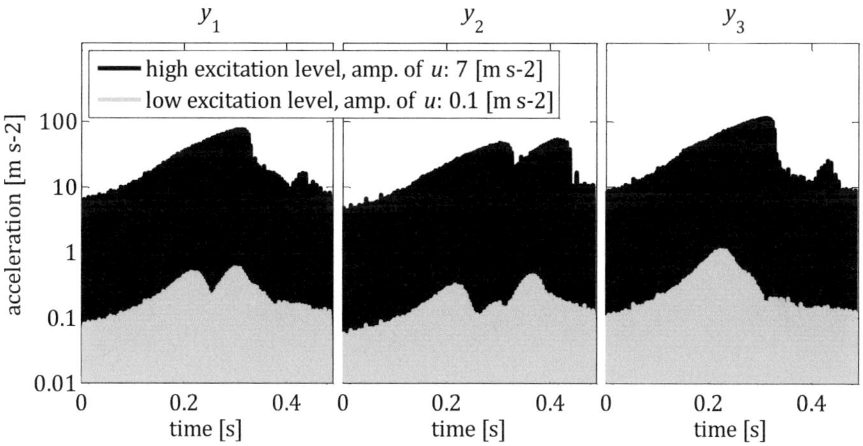

Figure 4.1: Noisy measurement signals of the virtual 3DoF system from Section 3.3.2 with logarithmic scaling in case of frequency sweep at two different excitation levels.

The global iteration procedure has been carried out using both the EKF and the SR-UKF, in order to compare their performance. The initial guess of the parameter vector has been set to $\mathbf{p}=[0,0,0.3\ \mathrm{kg}^{-1},0,0.3\ \mathrm{kg}^{-1}]^T$. The convergence of μ_{ij} over the number of global iterations is illustrated in Figure 4.2 and Figure 4.3 for the low level (linear) and the high level (nonlinear) load case respectively.

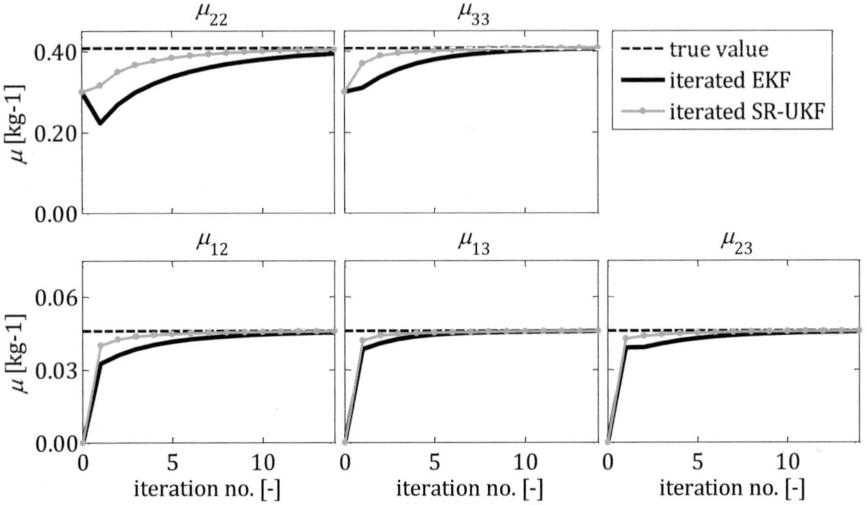

Figure 4.2: Convergence of the virtual 3DoF system's inertia properties from Section 3.3.2. The global iteration of the KF has been carried out based on the „linear" load case with low excitation level.

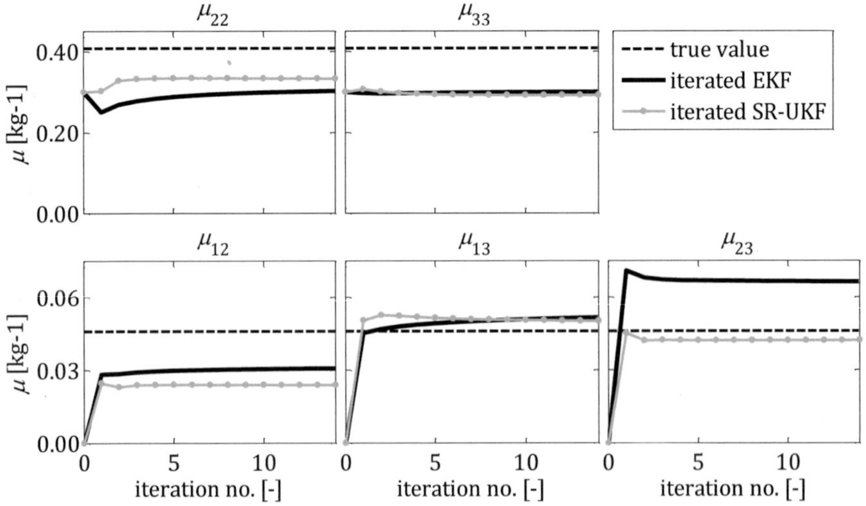

Figure 4.3: Convergence of the virtual 3DoF system's inertia properties from Section 3.3.2. The global iteration of the KF has been carried out based on the „nonlinear" load case with high excitation level.

First of all, it can be seen that the global iteration converges to stable results in 10 to 15 steps in all cases. The SR-UKF exhibits significantly faster convergence than the EKF, which can be explained by the trilinear nonlinearities in the model functions (q.v. Section 2.2.2). Furthermore, excellent estimation accuracy of the approach can be observed in case of low excitation level for both the EKF and the SR-UKF. Unfortunately this can not be stated about the original load case with high excitation level, where the converged parameters show significant deviation from their true values. For this measurement case the SR-UKF shows better average accuracy than the EKF. Therefore, the initial parameter guesses and the parameter estimates of the iterated SR-UKF from the load case with high excitation level has been used in a subsequent nonparametric RFS identification (according to Chapter 2) in order to assess, how critical the observed error of the identified constants really is. The comparison of the second spring element's RFCs, depicted in Figure 4.4, reveals that despite the seemingly large deviation of the estimated inertia properties a significant increase in the RFS identification quality could be achieved with the identified **M** compared to using its initial guess. Furthermore, the adapted model (involving the identified **M** and the identified RFSs) has been used to reconstruct the true system's response signals at high excitation level. Different zooms of these signals are compared to the true noisy (5 % RMS) measurement time series in Figure 4.5. A very good agreement between the depicted signals can be observed, which is a positive though rather unexpected result. A possible explanation to this phenomenon is that the high level excitation somehow weakens the distinguishability of the parameter vector's elements. Expanding the frequency range of the sweep did not reveal any hidden nonlinear resonances. Therefore, it can be concluded that for both low (linear) and high (nonlinear) excitation levels the globally iterated KF technique delivers adapted models that can reconstruct the true system's measured behaviour with good accuracy. If however the consistent identification of the model parameters is of high importance, then it is advised to carry out the iteration procedure based on low amplitude measurements.

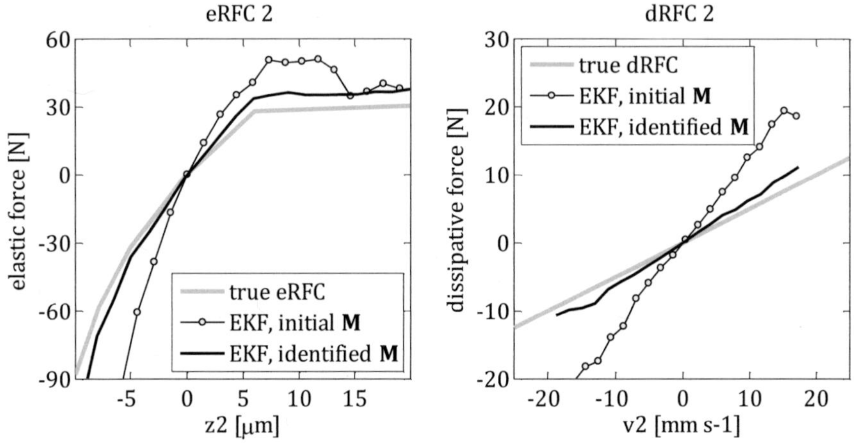

Figure 4.4: Identified eRFCs and dRFCs of the virtual 3DoF system's (q.v. Section 3.3.2) second spring element using the initial guess of **M** in comparison to using the identified **M** based on the high amplitude measurement with the iterated SR-UKF.

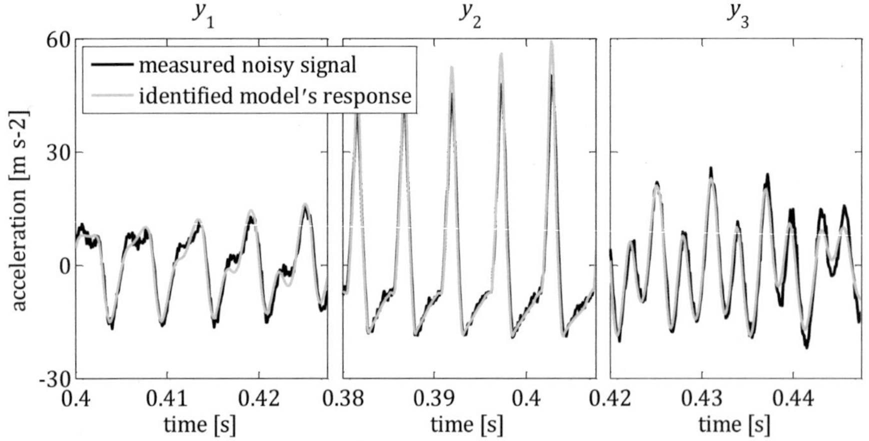

Figure 4.5: Noisy (5 % RMS) high amplitude response of the virtual 3DoF system (q.v. Section 3.3.2) compared to the adapted model's response using the identified RFSs and the estimated **M** based on the high amplitude measurement with the iterated SR-UKF.

Finally, it has to be mentioned that the convergence plots in Figure 4.2 and Figure 4.3 are slightly misleading regarding the computational performance of the EKF compared to the SR-UKF, since there is a significant difference in the CPU time of one iteration in case of the two different algorithms. Figure 4.6 aims to clarify this comparison using the normed CPU time (q.v. Section 3.5) of the iteration instead of the number of iteration steps. The estimation error curves of the inertia properties reveal that the EKF provides much faster convergence of the iteration than the SR-UKF. Hence, it turns out to be superior over the SR-UKF in the low excitation level case, where it exhibits the same estimation accuracy. Therefore, it can be concluded that the invested effort of keeping the model equations as weakly nonlinear as possible pays out both in result quality and in computational efficiency.

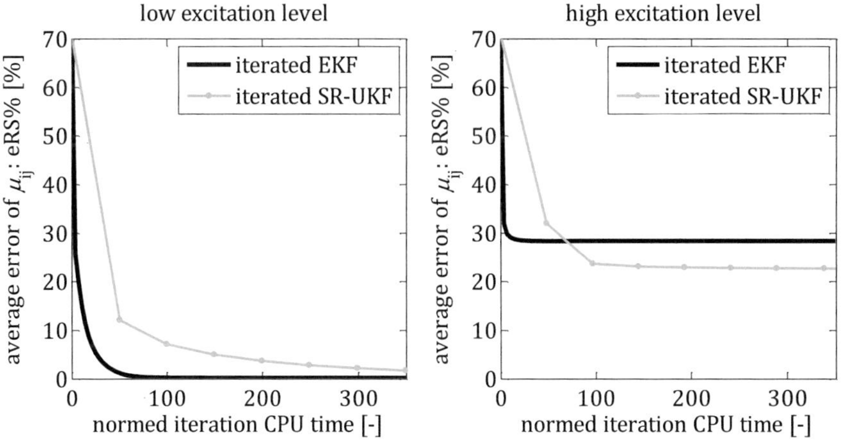

Figure 4.6: Average estimation error of the 3DoF system's (q.v. Section 3.3.2) inertia properties over the normed CPU time (q.v. Section 3.5) of the iterated EKF and the iterated SR-UKF based on the low amplitude (left) and the high amplitude (right) load case.

4.2 Estimation of the initial state

Recall from Section 3.3.2 that the proposed nonparametric identification approach yielded very good estimates of the 3DoF system's RFSs. This resulted in an adapted model that proved to be a good image of the true structure based on the comparison of their measurement responses. However, looking at Figure 3.20 in detail reveals significant deviation of the measurement signals in the initial phase of the frequency sweep. The reason for this is that the virtual system's true response begins with realistic steady state oscillatory conditions, just as in real-life, where one might start recording the measurement signals after switching on the excitation of the structure. The initial state of the adapted system however remained $[\mathbf{z}_0,\mathbf{v}_0]=[\mathbf{0},\mathbf{0}]$, since the KF only compensates but not adapts such initial errors. In this particular case this deviation of the adapted model can be neglected, since the influence of the initial state on the system's response vanishes relatively fast in case of stable mechanical structures with typical damping levels. However, in situations, where purely transient responses are measured or instable systems are to be identified, determining the proper values of $\mathbf{z}_0$ and $\mathbf{v}_0$ can gain notable importance.

A possible solution to this problem is the implementation of the Kalman Smoother (KS), whose special case, the Fixed-Interval RTS Smoother, has been presented in Section 1.6.3. It enhances the accuracy of the state estimates by involving not only past and present but also future measurement samples in the estimation of the state vector's time series. Since no past samples are available for the initial state, this extension of the KF is necessary to allow its adaption. In the current example the EKF in combination with the RTS Smoother, referred to as EKS, has been deployed for a short initial time period of the 3DoF systems high amplitude frequency response from Section 3.3.2. Similar to the case of the identification of $\mathbf{M}$ in Section 4.1, it turned out that the iteration of the algorithm with successively updated initial states leads to significant further increase of result quality, which is depicted in Figure 4.7. Thereby a very fast and accurate convergence of the initial velocity coordinates is to be observed. Though, the relative accuracy of the identified initial displacements is rather poor, it has to be mentioned that their absolute values were extremely small in the particular example.

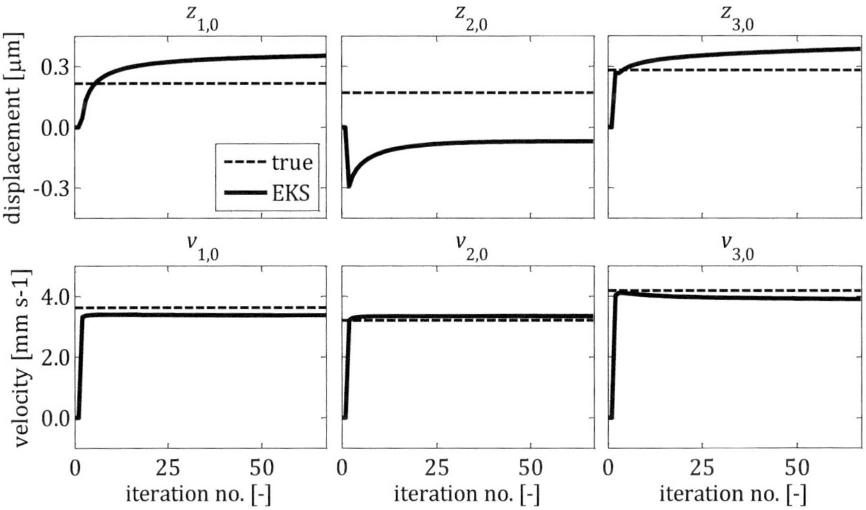

Figure 4.7: Convergence of the virtual 3DoF system's (q.v. Section 3.3.2) initial states based on the beginning interval of its noisy high amplitude measurement using the iterated EKS.

The initial time interval of the virtual system's measurement signals is compared to the adapted model's response in Figure 4.8. Apparently, the deployment of the iteratively identified initial state vector yields a drastic increase in the agreement of the depicted time series.

Finally, it is worth mentioning that the KS technique is dedicated to refine state variables, whose variation over time is symbolically described by the model equations. Therefore, it is not able to properly smooth states that are varying over time but assumed to be constant in the system model (Simon, 2006), such as **s** and **d**. This is the reason, why the exploratory implementation of the EKS in the proposed nonparametric identification approach did not show notable increase of accuracy. It could slightly refine the estimated time series of **z** and **v**, but unfortunately the main source of the identified RFS's bias lies in the finite synchronisation convergence of **s** and **d**.

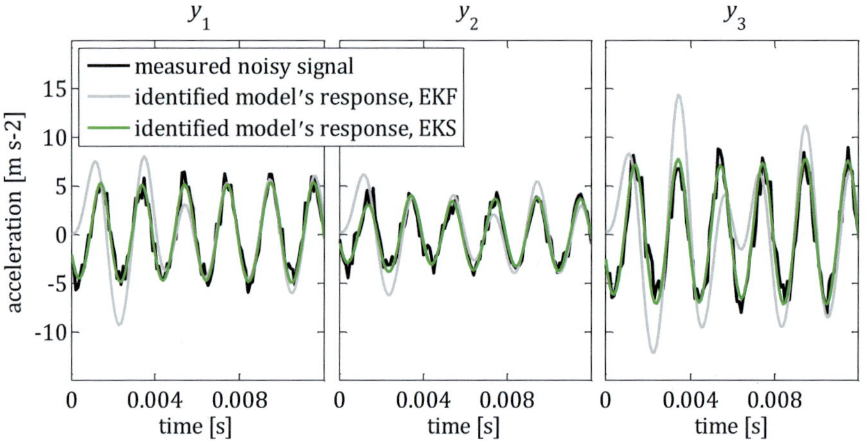

Figure 4.8: Initial time interval of the virtual 3DoF system's (q.v. Section 3.3.2) response signals. The true noisy time series (black) are compared to the adapted model's response involving initial states set to zero (grey) in comparison to initial states adapted using the iterated EKS (green).

5 Real-life implementation

The two identification case studies that are presented in this chapter investigate the complex properties of a Dual Mass Flywheel (DMF). A DMF is an important part of combustion engine drive trains. Its main purpose is the shifting of the system's rotational eigenfrequencies in order to avoid undesired noise phenomena such as gear box rattle or boom in the inside area of passenger cars. However, in order to produce DMFs of high quality, several further aspects, such as the axial dynamics of the flywheel, have to be taken into account as well. As the pioneer of the DMF, LuK GmbH & Co. KG, a member of the Schaeffler Group, has a great know-how in its design, simulation and production. This also includes sophisticated dynamic models that allow the consideration of the structure's behaviour under high level axial excitation (induced by the crank shaft of the combustion engine) in the early stage of product development. The experiments that are discussed in the following were carried out at LuK GmbH & Co. KG with the major purpose of determining damping ratios under realistic high level excitation in order to parametrise these simulation models. It is important to point out that the experiments were not specially designed to verify the proposed nonparametric identification approach. It has been deployed additionally to gain deeper insight into the structure's complex dynamics. This means that the test rig was not optimised to deliver ideal signals to support nonparametric identification, which proves the applicability of the approach under realistic conditions.

5.1 A 1DoF strongly nonlinear system

5.1.1 Experiment setup

The general structure of a DMF is illustrated in Figure 5.1 left. It mainly consists of the primary flywheel (PFW) and the secondary flywheel (SFW), which are coaxially linked together by a set of arc springs that define the rotational spring characteristic of the DMF. All further relative DoFs between the PFW and the SFW are blocked by a bearing, which is in the investigated particular design a compact combination of a plain bearing and an axial end-stop. The experiment test rig, depicted in Figure 5.1 right, consists of a high performance shaker platform, on which the PFW side of the DMF is rigidly mounted, which represents its designed mounting conditions. The arc springs have been rotationally prestressed, in order to attain realistic damping values. Both the applied excitation and the structure's response are recorded via accelerometers.

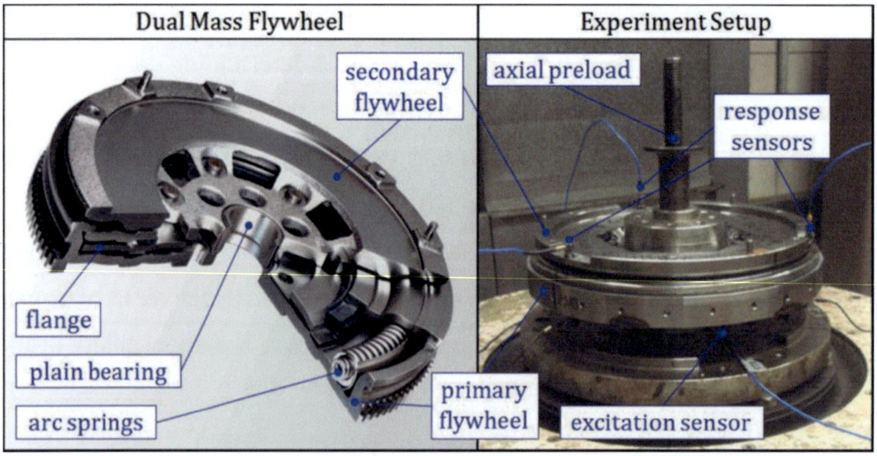

Figure 5.1: Basic components of a dual mass flywheel (left, source: http: //blog.motoringassist.com/ motoring-advice/wp-content/uploads/2013/12 /dual .jpg, Accessed: 3rd March 2016). The experiment setup on the shaker platform (right) with adjustable axial preload of the secondary flywheel.

The axial end-stop of the plain bearing acts only for pressing load. Therefore, it does not block against pulling the SFW and PFW apart. This is however not necessary, since under realistic operating conditions the SFW is always pressed against the PFW through the clutch system. To realise this condition, the test rig includes an adjustable axial preload unit, q.v. Figure 5.1 right. However, in order to generate an exotic load case, one can remove the axial preload, which allows the SFW to lift off from the PFW if a sufficient level of excitation is reached. This is exactly what has been carried out in the current example, which resulted in highly nonlinear behaviour of the system that is depicted in Figure 5.2.

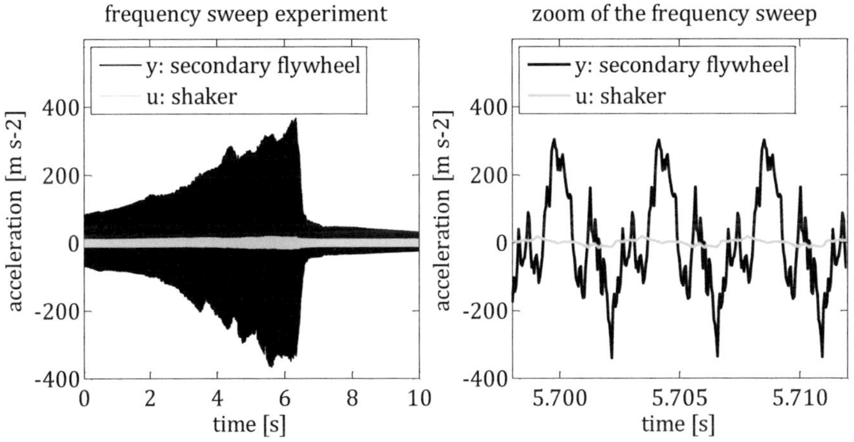

Figure 5.2: Measured axial vibration of the SFW during a forward frequency sweep from 200 Hz to 400 Hz without axial preload. Both u and y were measured via accelerometers. The signals were captured at a sampling rate of 20 kHz. Average RMS noise levels of 8 % and 19 % (*eRMS%*, q.v. Appendix B) have been determined for u and y respectively.

5.1.2 Nonparametric identification

The identification of the experiment setup, presented in Section 5.1.1, has been carried out using the proposed approach from Chapter 2. The vibration of the

SFW relative to the PFW is approximately one-dimensional due to the plain bearing. Therefore, its dynamics can be described by the 1DoF model, presented in Section 3.1. The only constant parameter of the system is the mass of the SFW. It can not be identified based the available measurement, since the excitation is given in form of acceleration. Therefore, the nominal mass of 4.96 kg has been used. The measured signals are highly corrupted by noise due to the impacts occurring at the end-stop of the plain bearing and at the other end of the SFW's axial backlash. This affects both the excitation u and the response signal y. Average RMS noise levels of 8 % and 19 % (*eRMS%*, q.v. Appendix B) have been determined for u and y respectively. Recall from Section 3.4.1 that the SR-UKF is superior over the EKF under such extreme conditions, which has been confirmed by the current application. First an additive identification of the structure's RFS has been carried out using the SR-UKF. The reconstructed RFCs are depicted in Figure 5.3. First of all, it can be seen that the system's characteristics are strongly nonlinear. Based on the a priori knowledge of the structure and the form of the measured response, the elastic RFC is expected to include backlash. Though the character of the identified eRFC is indeed progressive, the distribution of its samples and the high relative level of uncertainty for positive deformations reveal that the particular RFS can not be approximated by an additive model. Accordingly, the simulated response of the adapted additive system did not correlate with the measured response.

To overcome this problem, the derived coupled identification technique from Section 2.3.3 has been deployed. It led to a complex strongly nonlinear coupled RFS, which is depicted in Figure 5.4 left. The separation of single components of this complex characteristic is far from trivial. It represents the sum of the restoring forces that arise from the arc springs, the plain bearing and further contacts between the PFW and the SFW. Using this cRFS to reconstruct the measured oscillations showed a much better agreement with the true signal, q.v. Figure 5.4 right. It is well known that such nonlinear frequency responses exhibit multiple stable orbits at a given frequency. The collapse of a nonlinear resonance, i.e. the "jump" between these orbits, is extremely sensitive to damping. Accordingly, the identified cRFS has been slightly modified a posteriori via the application of an additional linear damping in order to tune the time point of the resonance's collapse. It is interesting to mention that due to noise disturbances in the measured excitation signal, this collapse tends to occur at certain impulse events in the virtual model. Therefore, it was not possible to perfectly tune this time point. To ensure that the adapted model's trajectory is initiated on the correct orbit, the iterated EKS (q.v. Section 4.2) has been used to identify proper entries of the initial state vector.

Compared to the virtual examples from Chapter 3, the current result in Figure 5.4 (right) might look unsatisfactory at first glance. However, this is an impression that is distorted by the high noise level of the true signal. Taking a look at the reconstructed time series in detail, depicted in Figure 5.5, reveals that the adapted system is indeed able to reconstruct the considered system's strongly nonlinear behaviour with good accuracy.

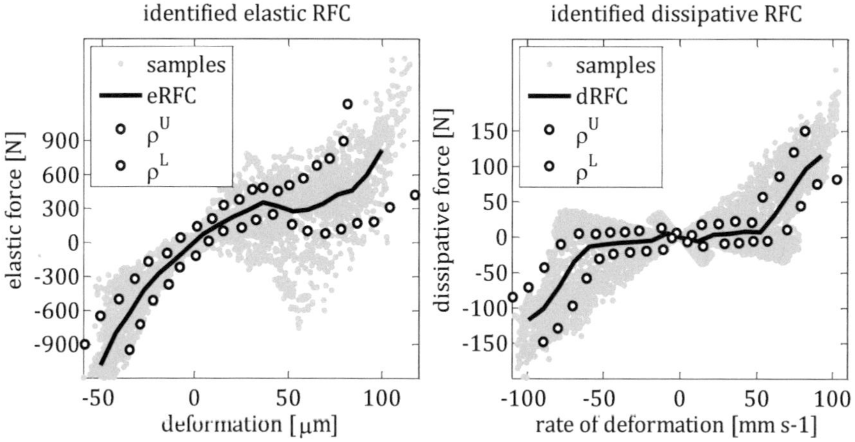

Figure 5.3: Identified eRFC (left) and dRFC (right) of the restoring force between the secondary flywheel and the shaker platform. $\boldsymbol{\rho}^U$ and $\boldsymbol{\rho}^L$ denote the estimated uncertainty curves according to Section 2.3.2.

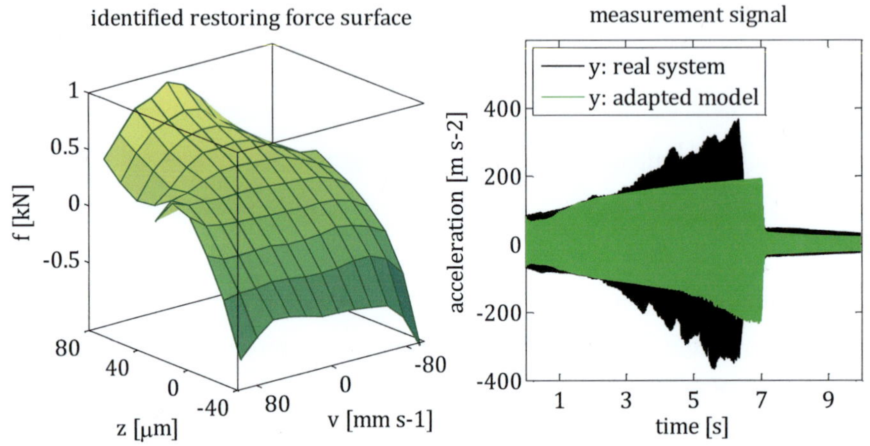

Figure 5.4: Identified cRFS of the secondary flywheel (left) and the adapted model's response time series compared to the real system's noisy measurement (right).

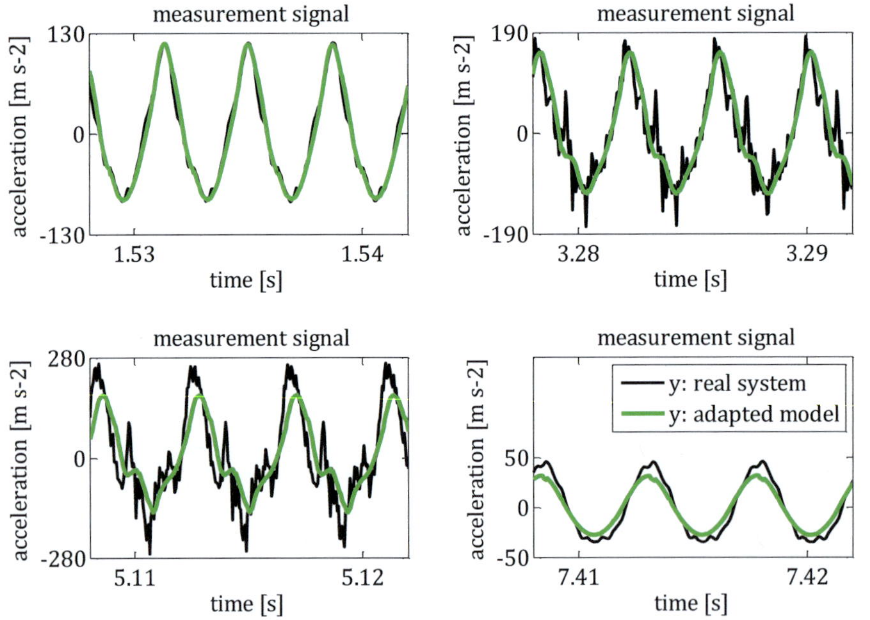

Figure 5.5: The adapted model's response time series compared to the real system's noisy measurement based on the zoom of the complete measurement at several time ranges.

5.2 A 3DoF weakly nonlinear system

5.2.1 Experiment setup

The 3DoF identification example is carried out based on the same experiment setup that was described in Section 5.1.1. However, in the current example realistic conditions are considered instead of an artificial load case. This means that the SFW is axially preloaded. This allows the undisturbed investigation of the PFW's properties. In contrast to the thick form of the cast iron SFW, the PFW is a rather thin structure formed from sheet steel. Its dynamics are mainly defined by its own elasticity, which typically leads to three elastic eigenmodes in the frequency range of approximately 100 Hz to 500 Hz. Three accelerometers on the outer diameter of the PFW that measure in axial direction are sufficient to distinguish these modes. Therefore, the experiment setup consists of the acceleration of the shaker platform as the input u, and the three response acceleration signals of the PFW, denoted by y_1, y_2 and y_3, at the angular positions 0°, 120° and 240° respectively in the $\{\xi,\eta\}$ plane, q.v. Figure 5.6.

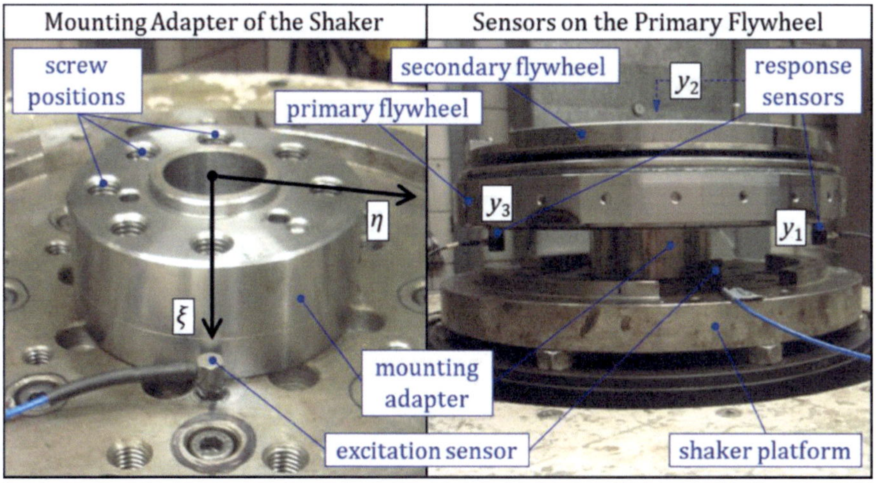

Figure 5.6: Experiment setup of the 3DoF identification case of the PFW.

The excitation signal and the system's three response signals are depicted in Figure 5.7 for the considered load case of a forward frequency sweep. Notice that the system seemingly exhibits only two resonances, although three eigenmodes are expected. The reason for this is not the improper frequency range of the sweep. This will be discussed in detail in Section 5.2.3.

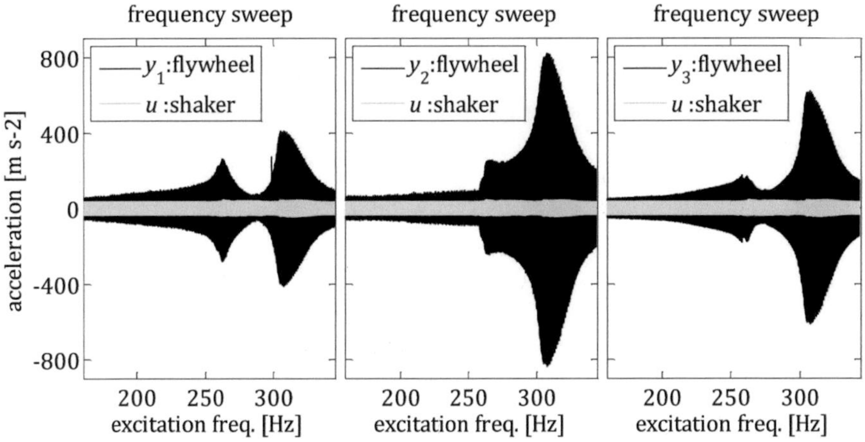

Figure 5.7: Measured axial vibration of the DMF's PFW during a forward frequency sweep with axially preloaded SFW. Both u and y_j were measured via accelerometers. The signals were captured at a sampling rate of 10 kHz. The time axis is replaced by the corresponding frequency of the excitation signal.

5.2.2 Full system identification

The current section presents the full identification of the PFW's given setup based on the frequency sweep measurement depicted in Figure 5.7. Although the observed dynamics are defined by elastic eigenmodes, it is shown in the following that the 3DoF rigid body model, which has been presented in Section 3.3.1 and been extended in Section 4.1, can be efficiently used to approximate the system. Therefore, the model given by Figure 3.18 is considered. According to Chapter 4 the first step is the identification of the structure's inertia properties using the iterated KF. The axial preload of the SFW led to a drastic reduction of the measured signals' noise level compared to Section 5.1.1. Accordingly, the computationally efficient EKF algorithm proved to be sufficient in the current case. The convergence of the inverse inertia matrix's entries over the iterations of the EKF is depicted in Figure 5.8. The initial values were chosen based on Finite Element simulations of the PFW. Notice that the choice of these inertia properties is not trivial, since the real structure exhibits continuum vibrations.

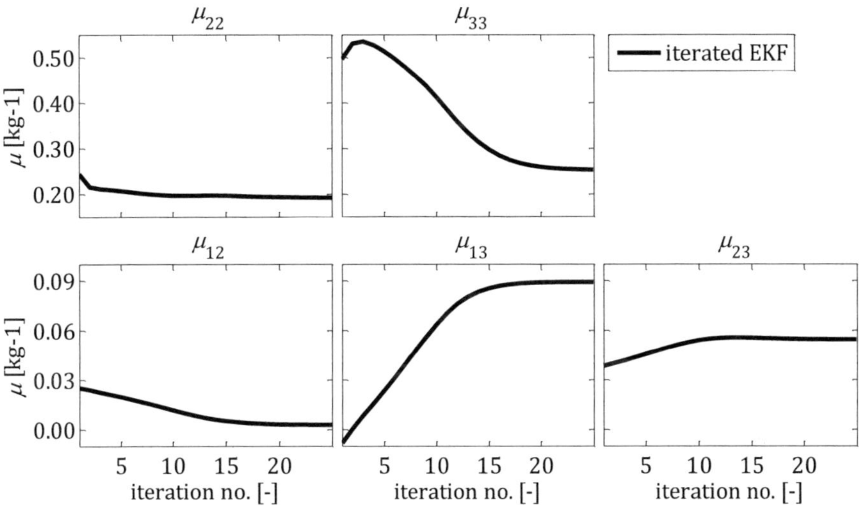

Figure 5.8: Convergence of the PFW's inertia properties over the iterations of the EKF.

Using the determined inverse inertia matrix $\mathbf{M}^{-1}$ a subsequential additive nonparametric identification of the model's three RFSs has been carried out using the EKF algorithm. The reconstructed elastic and dissipative RFCs are depicted in Figure 5.9 and Figure 5.10 respectively. It can be observed that the eRFCs of the system are approximately linear. This however can not be said about the dRFCs that reveal significant nonlinearities. Due to the high uncertainty and the extremely wide distribution range of dRFC1's samples, it could not be properly fitted using the presented automated algorithm. Deploying coupled fitting did not solve this problem, which indicates that the chosen a priori model probably neglects some dominant dissipative property of the real system at the first spring element, i.e. c_1 (q.v. Figure 3.18). Therefore, dRFC1 has been manually tuned in order to achieve the best possible agreement between the measured signals and the adapted model's response, which is depicted in Figure 5.11. Notice that the possibility of such separate characterisation of a multiple DoF system's RFSs is a major advantage of nonparametric identification.

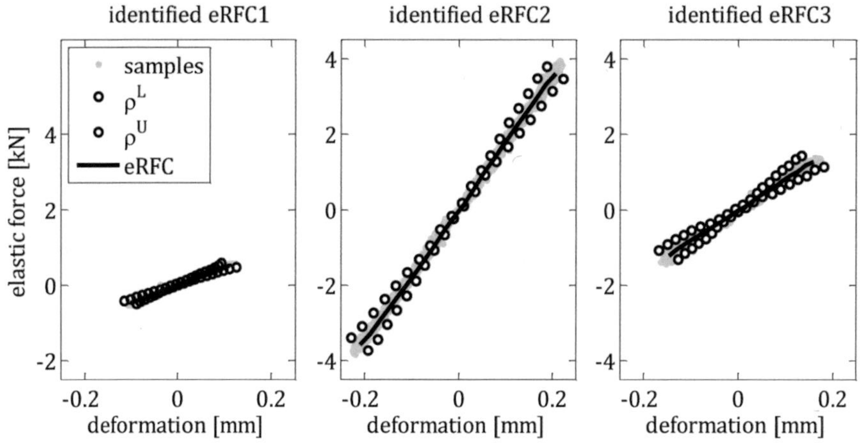

Figure 5.9: Identified elastic RFCs of the PFW's 3DoF model.

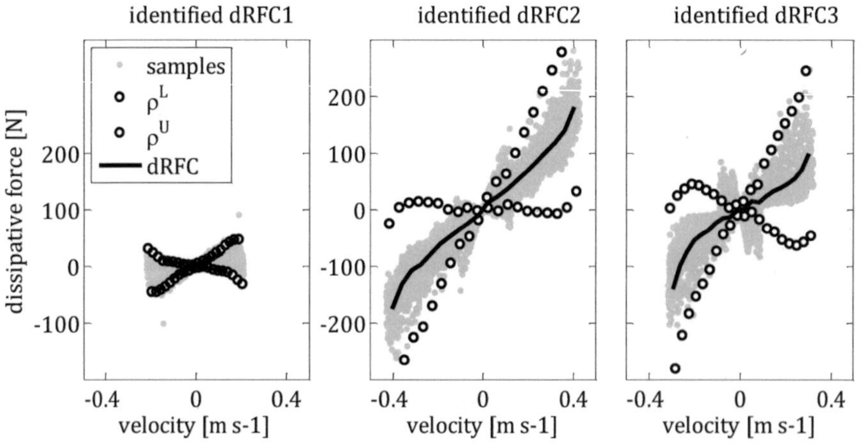

Figure 5.10: Identified dissipative RFCs of the PFW's 3DoF model.

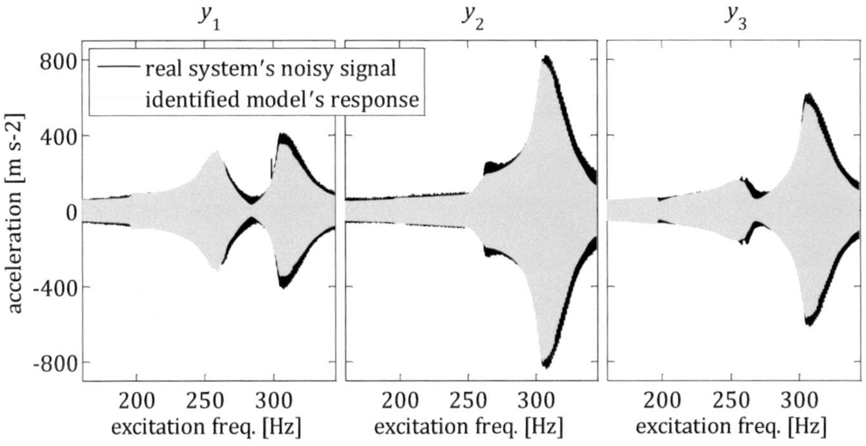

Figure 5.11: Comparison of the adapted 3DoF rigid body model's response to the measured continuum vibrations of the PFW during the frequency sweep experiment. The time axis is replaced by the corresponding frequency of the excitation signal.

5.2.3 Further system analysis

Nonlinearities are present in every engineering structure. Nevertheless, in case of weakly nonlinear systems, such as the PFW from Section 5.2.2, it is justified to identify linear models of the structure, which can approximately describe its behaviour in a particular load case. This allows the application of powerful linear analysis techniques such as Modal Analysis (MA). Though, this topic is not directly in the scope of the present thesis, it is briefly investigated based on the example of the PFW from Section 5.2.2. Thereby, the major aim is to highlight that the RFSs, which are identified by the proposed nonparametric approach, are not some kind of non-physical general functions that fit a model to a given measurement set. Instead they are nonparametric representations of physical properties that allow gaining insight into the measured system's structure.

Accordingly, the identified RFCs of the PFW have been replaced by constant stiffness and damping coefficients via the fitting of linear functions to the noisefree RFCs' nodes using simple Least Squares (LS) estimation. The calculation of the eigenvalues and eigenvectors of the first $2N\times2N$ elements of the a priori model's Jacobian (2.46) results in the three linearised eigenfrequencies and the corresponding mode shapes of the structure, which are depicted in Figure 5.12. First of all, it can be seen that the two tilting modes and the so called "potting mode", which are expected in case of such elastic plates, seem to be physically correct. Considering the modal displacement of the first tilting mode at the excitation position instantly explains, why this first mode was hard to see in the time signals of the frequency sweep measurement. It is now clear that the position of the excitation is near to the line of nonvibrating points of the particular mode shape. Furthermore, a relatively large difference between the frequencies of the two tilting modes can be observed, which is rather unexpected in case of such nearly rotational symmetric structures. Analysing the gradient lines of the mode shapes (indicated by red dashed lines in Figure 5.12) and reviewing the experiment setup leads to an obvious explanation of this phenomenon. Taking a look at the mounting of the DMF in Figure 5.6 (left), it can be observed that the seven mounting screws are not uniformly distributed. It can be seen in Figure 5.12 that the gradient line of the identified first tilting mode shape crosses between two screws, which are separated by a larger gap than the distance between the other screws. This explains the shifting of the first eigenfrequency to this unexpectedly low level, which is a result of the decreased mounting stiffness at the larger gap. It is easy to imagine that in case of a uniform mounting none of the tilting modes could be excited properly with the particular shaker platform, since it is only capable of axial excitation. However, the current non-uniform mounting allowed the observation of tilting vibrations as well. Though, it only led to the proper excitation of one tilting mode, thanks to the proposed identification algorithm both tilting eigenmodes of the system could be analysed.

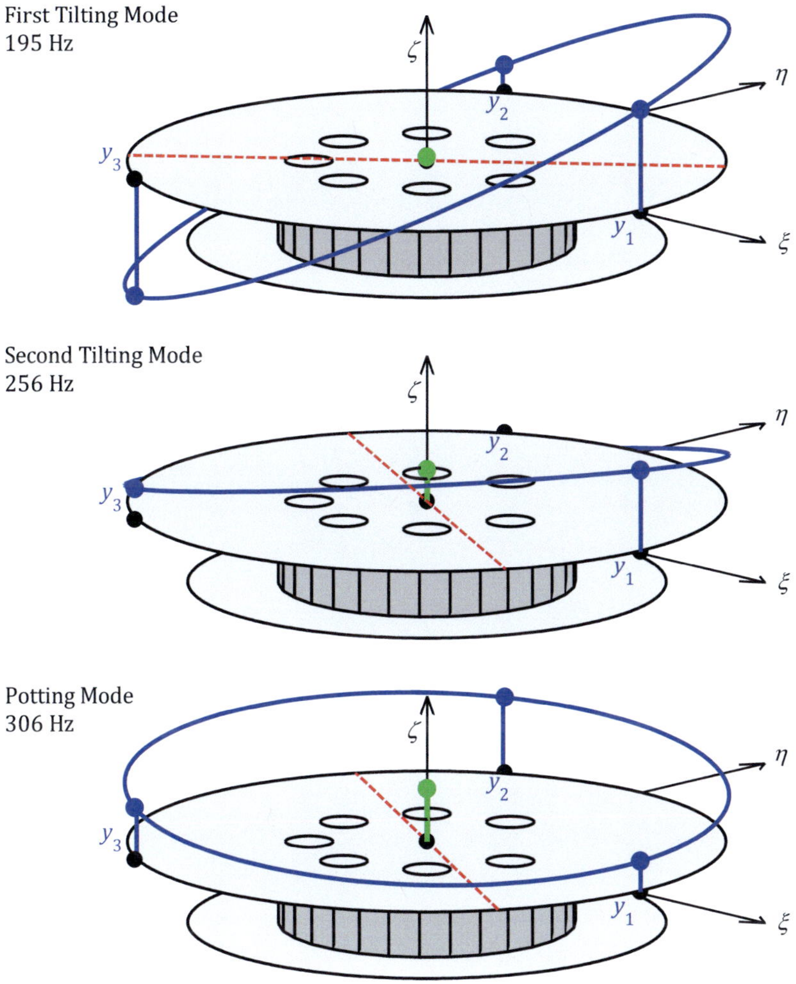

Figure 5.12: Eigenmodes of the measured primary flywheel based on the linearisation of its identified RFCs. Mode shapes (blue) are plotted including the modal displacement at the excitation position (green) and the gradient of the mode shape plane (red dashed line). The positions of the mounting screws are indicated by small black circles.

6 Conclusion

The present thesis explored the possibility of using Kalman Filter based master-slave synchronisation for the nonparametric identification of a subclass of strongly nonlinear dynamic systems, which, to the author's knowledge, has not yet been investigated in the literature. The core of the developed workflow consists of the synchronisation of a general nonparametric a priori model to the real system using noisy measurement signals to generate sample clouds of the system's restoring force characteristics. This is followed by optimal nonparametric fitting to generate noisefree elastic and dissipative Restoring Force Curves or coupled Restoring Force Surfaces. Additionally, the main algorithm has been coupled with well-known parametric identification techniques. As a result a full identification workflow of the defined subclass of systems has been proposed that allows the identification of their constant parameters as well. The detailed description of the algorithm was followed by several identification examples of one and three degree of freedom nonlinear systems, which allowed the extensive investigation of the derived technique's properties.

The following major advantages of the proposed approach have been revealed:

- It is nonparametric with respect to the system's restoring force characteristics, which therefore do not need to be characterised prior to the identification.
- It requires only one measured mechanical quantity per degree of freedom and provides identifiability using displacement, velocity and acceleration signals as well.
- The probabilistic manner of the procedure can accommodate signals highly corrupted by measurement noise without the application of frequency domain filters. Hence, it simultaneously filters the noise and preserves the high frequency nonlinear information.

- It provides a quantification of the identified restoring force characteristic's uncertainty.
- Due to the weakly nonlinear structure of the introduced general model equations, it allows the deployment of the fastest nonlinear Kalman Filter algorithm, the Extended Kalman Filter, under typical measurement conditions regardless of the type of nonlinearities that are to be identified. This results in a computationally efficient algorithm.
- It is a flexible approach due to the implementation of the Kalman Filter, which allows its extension by parametric methods to identify constant system properties as well.

Despite the above mentioned numerous advantages there are also some challenging aspects of the presented technique, which should be mentioned as well:

- A certain lower bound of the observable effective stiffness has been revealed, which introduces a restriction to the form of identifiable nonlinearities in case of velocity and acceleration measurements. Arbitrary restoring force surfaces can only be identified based on displacement signals.
- The approach is sensitive to modelling errors due to its synchronisation based nature. The additional parametric identification of uncertain model parameters has been investigated as a possible solution to overcome this difficulty.

On the whole it can be concluded that the investigated technique turned out to be a powerful approach for the identification of strongly nonlinear systems. Possible future work could consider the identification of structures with more than three degrees of freedom and could extend the considered class of dynamic systems.

Appendix

A Mathematical apparatus

The current appendix describes some of the mathematical formulas that are used in the present thesis.

Lie derivatives

The ith order Lie derivative of the scalar function $h(\mathbf{x})$ with respect to the vector function $\mathbf{a}(\mathbf{x})$ is denoted by $\mathrm{L}^i_{\mathbf{a}}(h)$, where $\mathbf{a}$ is a symbolic expression of the time derivative of $\mathbf{x}$, i.e. $\mathrm{d}\mathbf{x}/\mathrm{d}t=\mathbf{a}(\mathbf{x})$. It is defined by an implicit formula, where the 0th order derivative is $h(\mathbf{x})$ itself, i.e. $\mathrm{L}^0_{\mathbf{a}}(h):=h$. The further derivatives are given as

$$\mathrm{L}^{i+1}_{\mathbf{a}}(h) := \left(\nabla \mathrm{L}^i_{\mathbf{a}}(h)\right)^{\mathrm{T}} \mathbf{a}(\mathbf{x}) := \left(\frac{\partial \mathrm{L}^i_{\mathbf{a}}(h)}{\partial \mathbf{x}}\right)^{\mathrm{T}} \mathbf{a}(\mathbf{x}) \tag{A.1}$$

Trigonometric equalities

$$\sin^2(\varphi) + \cos^2(\varphi) = 1 \tag{A.2}$$

$$\sin^2(\varphi) = {}^1\!/_2 - {}^1\!/_2 \cos(2\varphi) \tag{A.3}$$

$$\cos^2(\varphi) = {}^1\!/_2 + {}^1\!/_2 \cos(2\varphi) \tag{A.4}$$

$$\sin(\varphi)\cos(\varphi) = {}^1\!/_2 \sin(2\varphi) \tag{A.5}$$

Statistical equalitics

Consider four scalar random variables A, B, C and D, which are described by their mean values a, b, c, d and their variances P_A, P_B, P_C and P_D respectively.

Their cross-covariances are given by P_{AB}, P_{AC}, P_{AD}, P_{BC}, P_{BD} and P_{CD}, i.e. their joint probability distribution is of the form

$$\begin{bmatrix} A \\ B \\ C \\ D \end{bmatrix} \sim \left(\begin{bmatrix} a \\ b \\ c \\ d \end{bmatrix}, \begin{bmatrix} P_A & P_{AB} & P_{AC} & P_{AD} \\ \cdot & P_B & P_{BC} & P_{BD} \\ \cdot & \cdot & P_C & P_{CD} \\ \cdot & \cdot & \cdot & P_D \end{bmatrix} \right). \tag{A.6}$$

If the joint distribution around the mean vector is symmetric, then according to (Simon, 2006) the following equalities hold:

$$\mathrm{E}(A-a)=0\,, \tag{A.7}$$

$$\mathrm{E}((A-a)^2)=P_A\,, \tag{A.8}$$

$$\mathrm{E}\big((A-a)(B-b)\big)=P_{AB}\,, \tag{A.9}$$

$$\mathrm{E}\big((A-a)(B-b)(C-c)\big)=\mathrm{E}\big((A-a)^2(B-b)\big)=\mathrm{E}((A-a)^3)=0\,, \tag{A.10}$$

where E(...) denotes the expectation operator. Furthermore, if the joint distribution is not only symmetric but also Gaussian, then according to (Gelb, Kasper, Nash, Price, & Sutherland, 2001) the following equalities are also true:

$$\mathrm{E}((A-a)^4)=3P_A^2\,, \tag{A.11}$$

$$\mathrm{E}((A-a)^2(B-b)^2)=P_AP_B+2P_{AB}^2\,, \tag{A.12}$$

$$\mathrm{E}\big((A-a)(B-b)(C-c)(D-d)\big)=P_{AB}P_{CD}+P_{AC}P_{BD}+P_{AD}P_{BC}\,. \tag{A.13}$$

B Error assessment

The comparison of the virtual model to the real system regarding its properties and dynamic behaviour is a frequently occurring task in system identification. The quantification of estimation accuracy, the convergence monitoring of the identification algorithm and the comparison of different identification techniques require appropriate quality factors. In the following the necessary error definitions are introduced, which are consistently used throughout the current thesis.

Consider two vector series $\mathbf{x}_i$ and $\mathbf{y}_i$, which are of the same dimension n_x. To quantify the error of $\mathbf{y}_i$ with respect to the reference series $\mathbf{x}_i$ on the interval $i\in\{i_0,i_0+1,\ldots,i_0+n_s-1\}$ the normalised Mean Square Error (eMS) is introduced in the following form:

$$eMS_{\mathbf{y},\mathbf{x}}^{i_0,n_s} := \frac{1}{n_x}\sum_{j=1}^{n_x}\left(\sum_{i=i_0}^{i_0+n_s-1}\left(x_{j,i}-y_{j,i}\right)^2\left[\sum_{i=i_0}^{i_0+n_s-1}\left(x_{j,i}-\bar{x}_j\right)^2\right]^{-1}\right), \tag{B.1}$$

where $\bar{x}_j$ denotes the average of $x_{j,i}$ over i. In most cases $\mathbf{y}_i$ is the estimate of $\mathbf{x}_i$, but it can also be e.g. a noise-corrupted version of $\mathbf{x}_i$. The simplified notation eMS is used without sub- and superscript if the latter are clear from the context. This definition defers from the simple mean square error in that it is normalised by the variance of the reference series, which enables the direct comparison of eMS values between completely different $\{\mathbf{x}_i,\mathbf{y}_i\}$ data pairs. Its relation to the Coefficient of Determination, denoted by R^2, which is commonly used in regression analysis (Yan & Su, 2009), is given by

$$eMS = 1 - R^2\,. \tag{B.2}$$

In order to directly compare the error to the analysed data, quality factors of compatible unit are defined in form of the normalised Root Mean Square Error (*eRMS*) and the normalised Percental Root Mean Square Error (*eRMS%*), which are given by the simple formula

$$eRMS\% := 100 \cdot eRMS := 100 \cdot \sqrt{eMS}\,. \tag{B.3}$$

For the scalar case, where y is compared to x, the above definitions reduce to the normalised Root Square Error (*eRS*) and the normalised Percental Root Square Error (*eRS%*), which are given as

$$eRS\% := 100 \cdot eRS := 100 \cdot \sqrt{(x-y)^2/x^2}\,. \tag{B.4}$$

Notice that *eRMS* is based on a quadratic error summation, and therefore it expresses both the bias (offset) and the variance (noisiness) of $\mathbf{y}_i$ compared to $\mathbf{x}_i$. In order to distinguish these two error components, an additional quality factor, the normalised Mean Error (*eM*), is introduced as

$$eM_{\mathbf{y},\mathbf{x}}^{i_0,n_s} := \frac{1}{n_x}\sum_{j=1}^{n_x}\left|\frac{1}{n_s}\sum_{i=i_0}^{i_0+n_s-1}\left(x_{j,i}-y_{j,i}\right)\left[\frac{1}{n_s}\sum_{i=i_0}^{i_0+n_s-1}\left(x_{j,i}-\bar{x}_j\right)^2\right]^{-\frac{1}{2}}\right|. \tag{B.5}$$

Since *eM* is based on linear error summation, it filters out symmetric error, i.e. variance. Therefore, the bigger the difference between *eRMS* and *eM*, the more dominant is the variance component of the error. The normalised Percental Mean Error (*eM%*) is defined as

$$eM\% := 100 \cdot eM\,. \tag{B.6}$$

C Observability of the exploratory system models

Finding an a priori process model that ensures observability for all kinematic measurement types is investigated in Section 2.1, which results in the RFS model $c(z,v){:=}sz{+}dv$. Thereby the properties of three additional exploratory models are discussed in Section 2.1.3 in order to justify the chosen model. The observability investigation of these additional models is summarised in the following. The first exploratory RFS model is of the form

$$c(z,v) := sz\,. \tag{C.1}$$

The corresponding observability matrices of the system for the three measurement types, displacement (z), velocity (v) and acceleration ($\dot{v}$), are:

$$\mathbf{O}|_{y=z} = \begin{bmatrix} 1 & 0 & -m^{-1}s \\ 0 & 1 & 0 \\ 0 & 0 & -m^{-1}z \end{bmatrix}, \tag{C.2}$$

$$\mathbf{O}|_{y=v} = \begin{bmatrix} 0 & -m^{-1}s & 0 \\ 1 & 0 & -m^{-1}s \\ 0 & -m^{-1}z & -m^{-1}v \end{bmatrix}, \tag{C.3}$$

$$\mathbf{O}|_{y=\dot{v}} = \begin{bmatrix} -m^{-1}s & 0 & m^{-2}s^2 \\ 0 & -m^{-1}s & 0 \\ -m^{-1}z & -m^{-1}v & m^{-2}sz + m^{-2}(sz - mbu) \end{bmatrix}. \tag{C.4}$$

Apparently all columns of these matrices are symbolically independent, which means that the model in general allows observability for all kinematic measurement types. However, it can be seen that symbolically zero rows appear for s=0 in case of $y \in \{v, \dot{v}\}$. Furthermore, it can be seen that for each measurement type the first and the third columns of $\mathbf{O}$ become linearly dependent if the particular measured quantity equals zero. In order to come to this conclusion for $y{=}\dot{v}$, it has to be recognised that $sz{-}mbu{=}{-}m\dot{v}$. Such situation occurs at least twice a period in case of typical vibration measurements, which leads to local observer divergencies for all three measurement types.

The second exploratory RFS model is of the form

$$c(z,v) \coloneqq sv\,. \tag{C.5}$$

The corresponding observability matrices of the system for the three measurement types, displacement (z), velocity (v) and acceleration ($\dot{v}$), are:

$$\mathbf{O}|_{y=z} = \begin{bmatrix} 1 & 0 & 0 \\ 0 & 1 & -m^{-1}s \\ 0 & 0 & -m^{-1}v \end{bmatrix}, \tag{C.6}$$

$$\mathbf{O}|_{y=v} = \begin{bmatrix} 0 & 0 & 0 \\ 1 & -m^{-1}s & m^{-2}s^2 \\ 0 & -m^{-1}v & m^{-2}(2sv - mbu) \end{bmatrix}, \tag{C.7}$$

$$\mathbf{O}|_{y=\dot{v}} = \begin{bmatrix} 0 & 0 & 0 \\ -m^{-1}s & m^{-2}s^2 & -m^{-3}s^3 \\ -m^{-1}v & m^{-2}(2sv - mbu) & m^{-3}(2msbu - 3s^2v) \end{bmatrix}, \tag{C.8}$$

The system is apparently not observable in case of $y \epsilon \{v, \dot{v}\}$, because the corresponding $\mathbf{O}$ matrices contain a zero row vector. This confirms the reasoning in Section 2.1.3, since the model (C.5) does not establish any symbolic connection between z and the measured quantity. Additionally in case of y=z the third row vector of $\mathbf{O}$ becomes zero for zero velocity, which similar to model (C.1) leads to local observer divergencies during vibration measurements.

The third exploratory RFS model is of the form

$$c(z,v) \coloneqq s(z^2 + \tau^2 v^2)\,, \tag{C.9}$$

where τ is a positive time constant that is necessary to match the units of z and v. Because of the complexity of the corresponding observability matrices, only their determinants are given in the following. As discussed in Section 2.1.2 these determinants are directly related to the observability criterion for single measurement systems. The symbolic determinants of **O** for the three measurement types, displacement (z), velocity (v) and acceleration ($\dot{v}$), are:

$$\det(\mathbf{O}|_{y=z}) = -m^{-1}[\, z^2 + \tau^2 v^2 \,] \tag{C.10}$$

$$\det(\mathbf{O}|_{y=v}) = -m^{-3} s v[\, 2z^2 m - \tau^2(2v^2 m - 4zmbu + 4sz^3 + 4s\tau^2 z v^2) \,] \tag{C.11}$$

$$\det(\mathbf{O}|_{y=\dot{v}}) = -m^{-6} s^2[\, \cdots \,] \tag{C.12}$$

It can be seen that similar to model (2.5) the determinant is symbolically zero for s=0 in case of $y\epsilon\{v, \dot{v}\}$. However, in this situation it is a major problem, since in the particular model s is not an effective stiffness but a direct coefficient of the restoring force, hence, it crosses zero and changes its sign at least twice a period for every typical oscillatory system, which results in local observer divergencies. Additionally in case of y=v the determinant becomes zero for v=0 as well. Nevertheless, for displacement measurement the specific model assures observability without restrictions, which however also holds for the model (2.5) that at the same time introduces a significantly weaker nonlinearity in the a priori process equation.

D Coefficients of the Unscented Transformation

Section 1.6.2 briefly describes the Unscented Transformation (UT) that is tuned by the coefficients $p_{i.j}^{\mathrm{UT}}$. According to (van der Merwe & Wan, 2001) these tuning coefficients are defined as follows.

$$p_1^{\mathrm{UT}} := \sqrt{(n + p_\lambda)} \tag{D.1}$$

$$p_\lambda := p_\alpha^2 (n + p_\kappa) - n \tag{D.2}$$

$$p_{2.j}^{\mathrm{UT}} := \frac{p_\lambda}{n + p_\lambda}, \qquad j = 1 \tag{D.3}$$

$$p_{3.j}^{\mathrm{UT}} := \frac{p_\lambda}{n + p_\lambda} - p_\alpha^2 + p_\beta + 1, \qquad j = 1 \tag{D.4}$$

$$p_{2.j}^{\mathrm{UT}} := p_{3.j}^{\mathrm{UT}} := \frac{1}{2(n + p_\lambda)}, \qquad j \epsilon \{2, \dots, 2n + 1\} \tag{D.5}$$

Thereby n denotes the size of the vector that is being transformed, which is the vector $\mathbf{x}$ in Section 1.6.2. The scaling factor p_α controls the spread of the sigma points and is to be chosen from the range of $p_\alpha \epsilon [10^{-4}, 1]$. The parameter p_β allows the incorporation of a priori knowledge about the distribution that is being transformed. For a Gaussian distribution $p_\beta = 2$ is the optimal choice. The secondary scaling factor p_κ is set to $p_\kappa = 0$ if the UT is used in a Kalman Filter (i.e. in the UKF). If the UT is deployed in recursive nonlinear parameter estimation, then it should be set to $p_\kappa = (3 - n)$.

E Central Moving Average

The Central Moving Average (CMA) is a simple off-line algorithm that generates a phase-shift-free moving average of the discretely sampled time series y, denoted by y^{CMA}. The averaged (smoothed) value for the jth sample is determined by the formula:

$$y_j^{\mathrm{CMA}k} := \frac{1}{k} \sum_{i=j-\frac{k-1}{2}}^{j+\frac{k-1}{2}} y_i \tag{E.1}$$

where k denotes the window size of the averaging, which is necessarily an odd integer. It should be set to at least 3, since k=1 results in the original signal. The "k-point" CMA of y is denoted by $y^{\mathrm{CMA}k}$. The first and last (k-1)/2 samples of the time series are smoothed using a reasonably reduced sample window.

This simple formulation implies an equidistant sampling over time, which is fulfiled in the examples throughout the current thesis. For a more general solution one can deploy MATLAB's "smooth" algorithm (MathWorks, www.mathworks.com), of which the CMA represents a simplified version.

F Derivation of the RFC sample formulas

The detailed derivation of the formulas for the mean vector and the covariance matrix of the probabilistic RFC samples, i.e. $\mathbf{x}_{\mathrm{RFC}}$ and $\mathbf{P}_{\mathrm{RFC}}$, as functions of $\mathbf{x}_C$ and $\mathbf{P}_C$ is presented in the following. This means the algebraic evaluation of the formulas (2.60) and (2.61) in order to achieve (2.62) and (2.63) respectively. The evaluation is based on the mathematical apparatus of statistical equalities from Appendix A. These rely on the general assumption that the considered probability distributions are symmetric. Furthermore, during the derivation of $\mathbf{P}_{\mathrm{RFC}}$, i.e. (2.63), formulas for fourth order stochastic moments are deployed, which imply the additional assumption that the considered distribution is Gaussian. The formula (2.60) for $\mathbf{x}_{\mathrm{RFC}}$ is the direct expected value of the Taylor series (2.59). Hence, no detailed derivation of (2.62) is necessary, since the statistical equalities can be applied directly without intermediate algebraic steps. Therefore, only the derivation of (2.63) from the formula (2.61) needs to be discussed, which results in the analytical expressions for $\mathbf{P}_{\mathrm{RFC}}$. The substitution of (2.59) and (2.62) into (2.61) leads to the following expressions:

$$P_{\mathrm{RFC},1,1} = (Z - z)^2 \qquad \mathbf{(F.1)}$$

$$P_{\mathrm{RFC},1,2} = (S - s)(Z - z)^2 + (S - s)(Z - z)z + (Z - z)^2 s - P_{ZS}(Z - z) \qquad \mathbf{(F.2)}$$

$$P_{\mathrm{RFC},1,3} = (Z - z)(V - v) \qquad \mathbf{(F.3)}$$

$$P_{\mathrm{RFC},1,4} = d(V - v)(Z - z) + (D - d)(V - v)(Z - z) + (D - d)(Z - z)v - P_{VD}(Z - z) \qquad \mathbf{(F.4)}$$

$$\begin{aligned} P_{\mathrm{RFC},2,2} = {} & (S-s)^2(Z-z)^2 + 2(S-s)^2(Z-z)z + (S-s)^2z^2 \\ & + 2(S-s)(Z-z)^2s + 2(S-s)(Z-z)sz + (Z-z)^2s^2 \\ & - 2P_{ZS}(S-s)(Z-z) - 2P_{ZS}(S-s)z - 2P_{ZS}(Z-z)s \\ & + P_{ZS}^2 \end{aligned} \tag{F.5}$$

$$\begin{aligned} P_{\mathrm{RFC},2,3} = {} & (S-s)(V-v)(Z-z) + (S-s)(V-v)z + (V-v)(Z-z)s \\ & - P_{ZS}(V-v) \end{aligned} \tag{F.6}$$

$$\begin{aligned} P_{\mathrm{RFC},2,4} = {} & d(S-s)(V-v)(Z-z) + d(S-s)(V-v)z + d(V-v)(Z \\ & - z)s + (D-d)(S-s)(V-v)(Z-z) \\ & + (D-d)(S-s)(V-v)z + (D-d)(S-s)(Z-z)v \\ & + (D-d)(S-s)vz + (D-d)(V-v)(Z-z)s \\ & + (D-d)(Z-z)sv - P_{VD}(S-s)(Z-z) - P_{VD}(S-s)z \\ & - P_{VD}(Z-z)s - P_{ZS}d(V-v) - P_{ZS}(D-d)(V-v) \\ & - P_{ZS}(D-d)v + P_{VD}P_{ZS} \end{aligned} \tag{F.7}$$

$$P_{\mathrm{RFC},3,3} = (V-v)^2 \tag{F.8}$$

$$P_{\mathrm{RFC},3,4} = d(V-v)^2 + (D-d)(V-v)^2 + (D-d)(V-v)v - P_{VD}(V-v) \tag{F.9}$$

$$\begin{aligned} P_{\mathrm{RFC},4,4} = {} & d^2(V-v)^2 + 2d(D-d)(V-v)^2 + 2d(D-d)(V-v)v \\ & + (D-d)^2(V-v)^2 + 2(D-d)^2(V-v)v + (D-d)^2v^2 \\ & - 2P_{VD}d(V-v) - 2P_{VD}(D-d)(V-v) - 2P_{VD}(D-d)v \\ & + P_{VD}^2 \end{aligned} \tag{F.10}$$

Applying the statistical equalities from Appendix A to the above expressions yields the following formulas for the entries of $\mathbf{P}_{\mathrm{RFC}}$, i.e. for (2.63):

$$P_{\mathrm{RFC},1,1} = P_Z \tag{F.11}$$

$$P_{\mathrm{RFC},1,2} = P_Z s + P_{ZS} z \tag{F.12}$$

$$P_{\mathrm{RFC},1,3} = P_{ZV} \tag{F.13}$$

$$P_{\mathrm{RFC},1,4} = P_{ZD} v + P_{ZV} d \tag{F.14}$$

$$P_{\mathrm{RFC},2,2} = P_S z^2 + P_Z s^2 + 2P_{ZS} sz + P_S P_Z - P_{ZS}^2 + 2P_{ZS}^2 \tag{F.15}$$

$$P_{\mathrm{RFC},2,3} = P_{VS} z + P_{ZV} s \tag{F.16}$$

$$P_{\mathrm{RFC},2,4} = P_{SD} vz + P_{VS} dz + P_{ZD} sv + P_{ZV} ds + P_{SD} P_{ZV} + P_{VS} P_{ZD} \tag{F.17}$$

$$P_{\mathrm{RFC},3,3} = P_V \tag{F.18}$$

$$P_{\mathrm{RFC},3,4} = P_V d + P_{VD} v \tag{F.19}$$

$$P_{\mathrm{RFC},4,4} = P_D v^2 + P_V d^2 + 2P_{VD} dv + P_D P_V - P_{VD}^2 + 2P_{VD}^2 \tag{F.20}$$

References

Amiri-Simkooei, A., & Jazaeri, S. (2012). *Weighted total least squares formulated by standard least squares theory.* Journal of Geodetic Science, DOI: 10.2478/v10156-011-0036-5.

Barnhart, B. L. (2011). *The Hilbert-Huang Transform: theory, applications, development.* Dissertation, University of Iowa, http://ir.uiowa.edu/etd/2670.

Bessa, W. M., Hackbarth, A., Kreuzer, E., & Radisch, C. (2014). *State and Parameter Estimation of an Electro-Hydraulic Servo System.* ENOC 2014, Vienna, Austria.

Carlsson, J., & Nordheim, C. (2011). *A Parameter Estimation Method for Continuous Time Dynamical Systems based on the Unscented Kalman Filter and Maximum Likelihood.* Master's Thesis, Chalmers University of Technology, Göteborg, Sweden.

Charbonneau, P. (2002). *An Introduction to genetic Algorithms for numerical Optimization.* NCAR, Colorado, Technical Note, NCAR/TN-450+IA.

Chatzi, E. N., & Smyth, A. W. (2009). *The Unscented Kalman Filter and Particle Filter methods for nonlinear structural system identification with non-collocated heterogeneous sensing.* Structural Control and Health Monitoring Vol.16.1, pp.99-123.

Chaurasiya, H. (2012). *Recent Trends of Measurement and Development of Vibration Sensors.* International Journal of Computer Science Issues, Vol.9, Iss.4, No.1.

Cox, H. (1964). *On the Estimation of State Variables and Parameters for Noisy Dynamic Systems.* IEEE Transactions on Automatic Control.

Crawley, E. F., & Aubert, A. C. (1986). *Identification of nonlinear structural elements by force-state mapping.* AIAA Journal, Vol.24, No.1, pp.155-162, DOI: 10.2514/3.9236.

Dresig, H., & Fidlin, A. (2014). *Schwingungen mechanischer Antriebssysteme.* Springer-Verlag Berlin Heidelberg, DOI: 10.1007/978-3-642-24117-8_1.

Feldman, M. (1994). *Non-linear System Vibration Analysis using Hilbert Transform - II. Forced Vibration Analysis Method 'FORCEVIB'.* Mechanical Systems and Signal Processing 8(3), pp.309-318.

Feldman, M. (1997). *Non-linear free vibration identification via the Hilbert transform.* Journal of Sound and Vibration 208(3), pp.475-489.

Fidlin, A. (2006). *Nonlinear Oscillations in Mechanical Engineering.* Springer Berlin Heidelberg New York, ISBN-13: 978-3-540-28115-3.

Fischer, A. (2013). *Implizite Runge-Kutta-Verfahren für die Zeitintegration steifer nichtlinearer mechanischer Systeme unter Berücksichtigung von Zwangsbedingungen.* Studienarbeit, Otto-von-Guericke-Universität Magdeburg, Fakultät für Maschinenbau, Institut für Mechanik.

Fox, J. (2002). *Nonparametric Regression.* Appendix to An R and S-PLUS Companion to Applied Regression.

Garcia, E., & Antsaklis, P. J. (2009). *Parameter Identification in Model Based Networked Control Systems Using Kalman Filters.* ISIS Technical Report ISIS-2009-004, June 2009.

Gelb, A., & Van der Velde, W. E. (1968). *Multiple-Input Describing Functions and Nonlinear System Design.* McGraw Hill.

Gelb, A., Kasper, J. F., Nash, R. A., Price, C. F., & Sutherland, A. A. (2001). *Applied Optimal Estimation.* The M.I.T. Press, ISBN: 0-262-20027-9.

Golub, G. H., & Van Loan, C. F. (1980). *An analysis of the total least squares problem.* SIAM Journal on Numerical Analysis, Vol.17, No.6, December 1980, pp.883-893.

Gunnarsson, G. S. (2014). *Identification of dynamical systems with a homotopy method: applications.* Bachelor Thesis, Karlsruhe Institute of Technology (KIT).

Hansen, P. C. (1992). *Analysis of Discrete Ill-Posed Problems by Means of the L-Curve.* SIAM Review, Vol.34, No.4, December 1992, pp.561-580.

Harris, C. M., & Piersol, A. G. (2002). *Harris' Shock and Vibration Handbook.* 5th Edition, McGraw-Hill, ISBN: 0-07-137081-1.

Hartikainen, J., Solin, A., & Särkkä, S. (2011). *Optimal filtering with Kalman filters and smoothers.* A manual for the Matlab Toolbox EKF/UKF, Version 1.3.

Hedrick, J. K., & Girard, A. (2013). *Control of Nonlinear Dynamic Systems: Theory and Applications.* http://www.me.berkeley.edu/ME237/ControlOfNonlinearDynamicSystems.pdf, Accessed: 3rd October 2013.

Hoshiya, M., & Saito, E. (1984). *Structural Identification by Extended Kalman Filter.* Journal of Engineering Mechanics 1984.110, pp.1757-1770.

Hu, H., & Proppe, C. (2011). *A Hilbert-Huang transform based identification method for general linear time-varying systems and weakly nonlinear systems.* Proceedings in Applied Mathematics and Mechanics 11, pp.227-228.

Hu, H., & Proppe, C. (2012). *Bayesian inference of linear time-varying systems based on Hilbert-Huang transform.* Proceedings in Applied Mathematics and Mechanics 12, pp.183–184, DOI: 10.1002/pamm.201210082.

Ibrahim, R. A. (2008). *Recent advances in nonlinear passive vibration isolators.* Journal of Sound and Vibration 314, pp.371-452, DOI: 10.1016/j.jsv.2008.01.014.

Ineichen, L. (2013). *Konzeptvergleich zur Bekämpfung der Torsionsschwingungen im Antriebsstrang eines Kraftfahrzeugs.* KIT Scientific Publishing, Karlsruhe, ISBN: 9783731500308.

Ing, J., Pavlovskaia, E., & Wiercigroch, M. (2011). *Complex Nonlinear Response of a Piecewise Linear Oscillator Experiment and Simulation.* Springer IUTAM Book Series 30, pp.135-143, ISBN: 978-94-007-1643-8_16.

Insperger, T., & Stépán, G. (2002). *Stability Chart for the Delayed Mathieu Equation.* Proceedings of Royal Society London A 458, pp.1989-1998.

Jakubek, S., & Fleck, A. (2009). *Inner Torque Estimation for Combustion Engines by Parametric Kalman Filtering.* Oldenbourg Wissenschaftsverlag, Automatisierungstechnik 57, 8, DOI: 10.1524/auto.2009.0786.

Julier, S. J., & Uhlmann, J. K. (1997). *A new extension of the Kalman filter to nonlinear systems.* Proc. SPIE Vol.3068, Signal Processing, Sensor Fusion and Target Recognition VI, pp.182-193.

Kalman, R. E. (1960). *A New Approach to Linear Filtering and Prediction Problems.* Transactions of the ASME–Journal of Basic Engineering, 82, Series D, pp.35-45.

Kalman, R. E., & Bucy, R. S. (1961). *New Results in Linear Filtering and Prediction Theory.* Journal of Basic Engineering, ASME, March 1961.

Kenderi, G., & Fidlin, A. (2014). *Nonparametric identification of nonlinear dynamic systems using a synchronization-based method.* Journal of Sound and Vibration 333, pp.6405-6423, DOI: 10.1016/j.jsv.2014.07.021.

Kerschen, G., Lenaerts, V., & Golinval, J. C. (2001). *Identification of wire rope isolators using the restoring force surface method.* International Conference on Structural System Identification, Kassel, pp.303-310.

Kerschen, G., Vakakis, A. F., Lee, Y. S., McFarland, D. M., & Bergman, L. A. (2008). *Toward a Fundamental Understanding of the Hilbert-Huang Transform in Nonlinear Structural Dynamics.* Journal of Vibration and Control 2008, 14, 77, DOI: 10.1177/1077546307079381.

Kerschen, G., Worden, K., Vakakis, A. F., & Golinval, J. C. (2006). *Past, present and future of nonlinear system identification in structural dynamics.* Mechanical Systems and Signal Processing 20, pp.505-592.

Kerschen, G., Worden, K., Vakakis, A. F., & Golinval, J. C. (2007). *Nonlinear System Identification in Structural Dynamics: Current Status and Future Directions.* SEM IMAC XXV, 25th International Modal Analysis Conference, Orlando, Florida.

Kiriakidis, K., & O'Brien Jr., R. T. (2004). *H-Infinity Parameter Estimation for State-Space Models.* Proceeding of 2004 American Control Conference, Boston, Massachusetts.

Kletschkowski, T. (2013). *Identification of Structural Parameters Based on Acoustic Measurements.* AIA-DAGA 2013 Conference on Acoustics, Merano.

Klöpper, R., Okuma, M., & Krüger, J. (2013). *Neues Verfahren zur Messung Vollständiger Trägheitseigenschaften.* MTZ, Motortechnische Zeitschrift, Vol.74, Iss.3, pp.240-245.

Kolansky, J., & Sandu, C. (2012). *Real-time parameter estimation study for inertia properties of ground vehicles.* The 2nd Joint International Conference on Multibody System Dynamics.

Kunath, S., Niemeier, R., Schlegel, R., & Will, J. (2014). *Kalibrierung von Messdaten und Simulationen als eine Optimierungsaufgabe mit Signalen.* NAFEMS Magazin 3/2014.

Lagarias, J. C., Reeds, J. A., Wright, M. H., & Wright, P. E. (1998). *Convergence Properties of the Nelder-Mead Simplex Method in Low Dimensions.* SIAM Journal of Optimization, Vol.9, No.1, pp.112-147.

Laxalde, D., & Thouverez, F. (2009). *Complex Non-Linear Modal Analysis for Mechanical Systems: Application to Turbomachinery Bladings With Friction Interfaces.* Journal of Sound and Vibration 322, 4-5, pp.1009-1025, DOI: 10.1016/j.jsv.2008.11.044.

Lenaerts, V., Kerschen, G., Golinval, J. C., Ruzzene, M., & Giorcelli, E. (2004). *Validation of two nonlinear system identification techniques using an experimental testbed.* Shock and Vibration 11, Section 3.5, pp.365-375.

Li, S. J., Suzuki, Y., & Noori, M. (2004). *Identification of hysteretic systems with slip using bootstrap filter.* Mechanical Systems and Signal Processing, Vol.18, Iss.4, pp.781-795, DOI: 10.1016/j.ymssp.2003.08.001.

Link, M., Boeswald, M., Laborde, S., Weiland, M., & Calvi, A. (2011). *Non-linear Experimental Modal Analysis and Application to Satellite Vibration Test Data.* III ECCOMAS Thematic Conference on Computational Methods in Structural Dynamics and Earthquake Engineering, Corfu, Greece.

Ljung, L. (2010). *Approaches to Identification of Nonlinear Systems.* 29th Chinese Control Conference, 29-31 July 2010, ISBN: 978-1-4244-6263-6.

Ljung, L. (2013). *Matlab System Identification Toolbox User's Guide, Version 6.* http://www.mathworks.com/help/releases/R13sp2/pdf_doc/ident/ident.pdf, Accessed: 22 October 2013.

Lourens, E., Papadimitriou, C., Gillijns, S., Reynders, E., De Roeck, G., & Lombaert, G. (2012). *Joint input-response estimation for structural systems based on reduced-order models and vibration data from a limited number of sensors.* Mechanical Systems and Signal Processing 29, pp.310-327.

Lourens, E., Reynders, E., De Roeck, G., Degrande, G., & Lombaert, G. (2012). *An augmented Kalman filter for force identification in structural dynamics.* Mechanical Systems and Signal Processing 27, pp.446-460.

Maplesoft. (www.maplesoft.com). *Maple.* Maplesoft, a division of Waterloo Maple Inc., 1996-2015.

Markovsky, I., & Van Huffel, S. (2007). *Overview of total least squares methods.* Signal processing 87, No.10, pp.2283-2302.

Masri, S. F., Chassiakos, A. G., & Caughey, T. K. (1992). *Structure-unknown non-linear dynamic systems: identification through neural networks.* Smart Materials and Structures Vol.1, pp.45-56.

MathWorks. (www.mathworks.com). *MATLAB.* The MathWorks Inc., 1994-2015.

McBurney, P. W. (1990). *A robust approach to reliable real-time Kalman filtering.* Position Location and Navigation Symposium, IEEE PLANS '90, Las Vegas, DOI: 10.1109/PLANS.1990.66227.

Meyer, M. J. (2003). *Monte Carlo Simulation with Java and C++.* http://www.javaquant.net/books/MCBook-1.2.pdf, Accessed: 25 September 2015.

Michalik, C., Hannemann, R., & Marquardt, W. (2009). *Incremental Single Shooting - A Robust Method for the Estimation of Parameters in Dynamical Systems.* Computers & Chemical Engineering, Vol.33, Iss.7, 15 July 2009, pp.1298-1305.

Mohamed, M., Karim, T., & Belghith, S. (2013). *Chaos Suppression in Forced Van Der Pol Oscillator.* International Journal of Computer Applications, 0975-8887, Vol.68, No.23.

Moreno, V. M., & Pigazo, A. (2009). *Kalman Filter: Recent Advances and Applications.* In-Tech, Croatia, ISBN: 978-953-307-000-1.

Németh, J. G., Kollár, I., & Schoukens, J. (2001). *Identification of Volterra Kernels Using Interpolation.* IEEE Instrumentation and Measurement Technology Conference, Budapest, Hungary, 21-23 May 2001.

Nicklas, R. B. (1989). *An Application of a Kalman filter fixed Interval Smoothing Algorithm to Underwater Target Tracking.* Naval Postgraduate School, Monterey, California, Master's Thesis.

Nijmeijer, H., & Angeles, A. R. (2003). *Synchronization of mechanical systems.* World Scientific, New York, ISBN: 9789812386052.

Nijmeijer, H., & Mareels, I. M. (1997). *An observer looks at synchronization.* IEEE Transactions on Circuits and Systems - I: Fundamental Theory and Applications, Vol.44, No.10.

Nocedal, J., & Wright, S. J. (1999). *Numerical Optimization.* Springer-Verlag New York, ISBN: 0-387-98793-2.

Noël, J. P., & Kerschen, G. (2013). *Nonparametric subspace identification of nonlinear structures - Application to a spacecraft.* 54th AIAA/ASME/ASCE/AHS/ASC Structures, Structural Dynamics and Materials Conference.

Noël, J. P., Kerschen, G., & Newerla, A. (2012). *Application of the Restoring Force Surface Method to a Real-life Spacecraft Structure.* Topics in Nonlinear Dynamics, Vol.3, Conference Proceedings of the Society for Experimental Mechanics Series 2012, pp.1-19.

Okuma, M., & Oho, T. (1997). *Experimental Spatial Matrix Identification Method.* 5th International Congress on Sound and Vibration, University of Adelaide, South Australia.

Park, Y. S., & Kim, W. J. (1994). *Experimental Identification of Coulomb Friction Parameter by Applying Force-State Mapping Technique in Frequency Domain.* SEM IMAC XII, 12th International Modal Analysis Conference.

Peeters, M. (2010). *Theoretical and Experimental Modal Analysis of Nonlinear Vibrating Structures using Nonlinear Normal Modes.* PhD Thesis, University of Liège.

Peifer, M., Timmer, J., & Voss, H. U. (2003). *Nonparametric identifcation of nonlinear oscillating systems.* Journal of Sound and Vibration 267, pp.1157-1167.

Platten, M. F., Wright, J. R., Cooper, J. E., & Sarmast, M. (2002). *Identification of Multi-degree of Freedom Non-linear Systems using an Extension of Force Appropriation.* IMAC-XX: Conference & Exposition on Structural Dynamics.

Prahs, A. (2011). *Implementierung von Methoden der nichtlinearen Zeitreihenanalyse.* Karlsruhe Institute of Technology, Bachelor Thesis.

Pulido Herrera, E., & Kaufmann, H. (2010). *Adaptive Methods of Kalman Filtering for Personal Positioning Systems.* 23rd International Technical Meeting of the Satellite Division of The Institute of Navigation, Portland.

Richardson, M. H. (1977). *Derivation of Mass, Stiffness and Damping Parameters From Experimental Modal Data.* Hewlett Packard Company, Santa Clara Division.

Santoboni, G., Pogromsky, A. Y., & Nijmeijer, H. (2003). *Partial Observers and Partial Synchronization.* International Journal of Bifurcation and Chaos, Vol.13, No.2, pp.453-458.

Schlecht, B. (2009). *Maschinenelemente 2: Getriebe, Verzahnungen und Lagerungen.* Pearson Deutschland GmbH, München, ISBN: 978-3-8273-7146-1.

Simon, D. (2001). *From Here to Infinity.* Embedded Systems Programming.

Simon, D. (2006). *Optimal State Estimation: Kalman, Hinf and Nonlinear Approaches.* John Wiley & Sons Inc.

Simon, D., & Simon, D. L. (2006). *Constrained Kalman Filtering via Density Function Truncation for Turbofan Engine Health Estimation.* NASA/TM-2006-214129, Glenn Research Center, Cleveland, Ohio 44135.

Sitz, A., Schwarz, U., Kurths, J., & Voss, H. U. (2002). *Estimation of parameters and unobserved components for nonlinear systems from noisy time series.* Physical Review E 66, 016210, DOI: 10.1103/PhysRevE.66.016210.

Stamm, W. (2011). *Modellierung und Simulation von Mehrkörpersystemen mit flächigen Reibkontakten.* Karlsruher Institut für Technologie, Schriftenreihe des Instituts für Technische Mechanik, Band 13.

Stépán, G. (1989). *Retarded Dynamical Systems, Stability and characteristic Functions.* Longman Scientific & Technical.

Stépán, G., Szalai, R., & Hogan, S. J. (2005). *The Chaotic Oscillations of High-Speed Milling.* IUTAM Symposium on Chaotic Dynamics and Control of Systems and Processes in Mechanics, Solid Mechanics and its Applications, Vol.122, pp.147-158.

Strogatz, S. H. (1994). *Nonlinear Dynamics and Chaos.* Perseus Books Publishing, L.L.C., ISBN: 0-201-54344-3.

Sturm, M., Moorhouse, A., Kropp, W., & Alber, T. (2013). *Robust force identification for complex technical structures with single degree of freedom excitation using an adaptive algorithm in time domain.* AIA-DAGA 2013, Merano.

Sun, Z., & Yang, Z. (2010). *Study of nonlinear parameter identification using UKF and maximum likelihood method.* Control Applications, IEEE International Conference, pp.671-676.

Tikhomolov, A. (2015). *Analytische, numerische und messtechnische Untersuchung der Dynamik von Fahrzeugkupplungen am Beispiel des Trennproblems.* KIT Scientific Publishing, Karlsruhe, ISBN: 9783731503620.

van der Merwe, R., & Wan, E. A. (2001). *The square-root unscented Kalman filter for state and parameter-estimation.* IEEE International Conference on Acoustics, Speech and Signal Processing, Vol.6, pp.3461-3464.

van der Merwe, R., & Wan, E. A. (2004). *Sigma-Point Kalman Filters for Integrated Navigation.* Institute of Navigation Annual Technical Meeting.

Viguie, R., & Kerschen, G. (2010). *On the functional form of a nonlinear vibration absorber.* Journal of Sound and Vibration 329, pp.5225-5232, DOI: 10.1016/j.jsv.2010.07.004.

Volkova, V. E. (2010). *Non-parametric Identification of Dynamic Model of Cracked Steel Reinforced Beam.* Proceedings of The 10th International Conference on Modern Building Materials, Structures and Techniques, Vilnius, Lithuania, pp.1066-1069.

Volkova, V. E. (2011). *Non-parametric identification of chaotic systems.* CMM-2011, Computer Methods in Mechanics, 9-12 May 2011, Warsaw, Poland.

Volkova, V. E. (2013). *Development of Methods for Nonparametric Identification of Models of Mechanical Systems.* Procedia Engineering 57, pp.1230-1235, DOI: 10.1016/j.proeng.2013.04.155.

Voss, H. U., & Kurths, J. (1997). *Reconstruction of Nonlinear Time Delay Models from Data by the Use of Optimal Transformations.* Physics Letters A 234, pp.336-344.

Voss, H. U., Rust, H., Horbelt, W., & Timmer, J. (2003). *A combined approach for the identification of continuous nonlinear systems.* International Journal of Adaptive Control and Signal Processing 17, pp.335-352.

Voss, H. U., Timmer, J., & Kurths, J. (2004). *Nonlinear dynamical system identification from uncertain and indirect measurements.* International Journal of Bifurcation and Chaos, Vol.14, No.6, pp.1905-1933.

Vyasarayani, C. P., Uchida, T., Carvalho, A., & McPhee, J. (2012). *Parameter Identification in Dynamic Systems Using the Homotopy Optimization Approach.* Springer-Verlag London, Identification for Automotive Systems, LNCIS 418, pp.129-145.

Walter, E. (2014). *Numerical Methods and Optimization, A Consumer Guide.* Springer International Publishing, ISBN: 978-3-319-07671-3, DOI: 10.1007/978-3-319-07671-3.

Wang, L., Gawthrop, P. J., & Young, P. C. (2005). *Continuous Time System Identification of Nonparametric Models with Constraints.* 16th IFAC World Congress, 4-8 July 2005, Prague.

Worden, K., & Tomlinson, G. R. (2001). *Nonlinearity in experimental modal analysis.* Philosophical Transactions of the Royal Society, A 2001 359, pp.113-130, DOI: 10.1098/rsta.2000.0716.

Wu, C., Lin, P. P., Han, Z., & Li, S. (2010). *Simulation-based optimal design of [alpha]-[beta]-[gamma]-[delta] filter.* International Journal of Automation and Computing 7.2, May 2010, pp.247-253.

Wu, M., & Smyth, A. W. (2007). *Application of the Unscented Kalman Filter for real-time nonlinear structural system identification.* Structural Control and Health Monitoring, Vol.14.7, pp.971-990.

Yan, X., & Su, X. G. (2009). *Linear Regression Analysis, Theory and Computing.* World Scientific, ISBN-13: 978-981-283-410-2.

Young, P. C. (2000). *Stochastic Dynamic Modelling and Signal Processing: Time Variable and State Dependent Parameter Estimation.* Nonlinear and Nonstationary Signal Processing, Cambridge University Press, Cambridge, pp.74-114.

Young, P. C., & Garnier, H. (2006). *Identification and Estimation of Continuous-Time, Data-Based Mechanistic (DBM) Models for Environmental Systems.* Environmental Modelling & Software 21, No.8, pp.1055-1072, DOI: 10.1016/j.envsoft.2005.05.007.

Zanotti Fragonara, L., Ceravolo, R., Matta, E., Quattrone, A., De Stefano, A., & Pecorelli, M. (2012). *Non-linear characterisation of the physical model of an ancient masonry bridge.* Journal of Physics: Conference Series 382, DOI: 10.1088/1742-6596/382/1/012051.

Zeile, C., & Maione, I. A. (2015). *Experimental results and validation of a method to reconstruct forces on the ITER test blanket modules.* Fusion Engineering and Design, DOI: 10.1016/j.fusengdes.2015.04.028.

Zhang, L., Gu, T., Zhao, J., Ji, S., Hu, M., & Li, X. (2013). *An Improved Moving Least Squares Method for Curve and Surface Fitting.* Hindawi Publishing Corporation, Mathematical Problems in Engineering, Article ID: 159694, http://dx.doi.org/10.1155/2013/159694.

KARLSRUHER INSTITUT FÜR TECHNOLOGIE (KIT)
SCHRIFTENREIHE DES INSTITUTS FÜR TECHNISCHE MECHANIK (ITM)

ISSN 1614-3914

Band 1 **Marcus Simon**
Zur Stabilität dynamischer Systeme mit stochastischer Anregung. 2004
ISBN 3-937300-13-9

Band 2 **Clemens Reitze**
Closed Loop, Entwicklungsplattform für mechatronische Fahrdynamikregelsysteme. 2004
ISBN 3-937300-19-8

Band 3 **Martin Georg Cichon**
Zum Einfluß stochastischer Anregungen auf mechanische Systeme. 2006
ISBN 3-86644-003-0

Band 4 **Rainer Keppler**
Zur Modellierung und Simulation von Mehrkörpersystemen unter Berücksichtigung von Greifkontakt bei Robotern. 2007
ISBN 978-3-86644-092-0

Band 5 **Bernd Waltersberger**
Strukturdynamik mit ein- und zweiseitigen Bindungen aufgrund reibungsbehafteter Kontakte. 2007
ISBN 978-3-86644-153-8

Band 6 **Rüdiger Benz**
Fahrzeugsimulation zur Zuverlässigkeitsabsicherung von karosseriefesten Kfz-Komponenten. 2008
ISBN 978-3-86644-197-2

Band 7 **Pierre Barthels**
Zur Modellierung, dynamischen Simulation und Schwingungsunterdrückung bei nichtglatten, zeitvarianten Balkensystemen. 2008
ISBN 978-3-86644-217-7

Band 8 **Hartmut Hetzler**
Zur Stabilität von Systemen bewegter Kontinua mit Reibkontakten am Beispiel des Bremsenquietschens. 2008
ISBN 978-3-86644-229-0

Band 9 **Frank Dienerowitz**
Der Helixaktor – Zum Konzept eines vorverwundenen Biegeaktors. 2008
ISBN 978-3-86644-232-0

Band 10 **Christian Rudolf**
Piezoelektrische Self-sensing-Aktoren zur Korrektur statischer Verlagerungen. 2008
ISBN 978-3-86644-267-2

Band 11 **Günther Stelzner**
Zur Modellierung und Simulation biomechanischer Mehrkörpersysteme. 2009
ISBN 978-3-86644-340-2

Band 12 **Christian Wetzel**
Zur probabilistischen Betrachtung von Schienen- und Kraftfahrzeugsystemen unter zufälliger Windanregung. 2010
ISBN 978-3-86644-444-7

Band 13 **Wolfgang Stamm**
Modellierung und Simulation von Mehrkörpersystemen mit flächigen Reibkontakten. 2011
ISBN 978-3-86644-605-2

Band 14 **Felix Fritz**
Modellierung von Wälzlagern als generische Maschinenelemente einer Mehrkörpersimulation. 2011
ISBN 978-3-86644-667-0

Band 15 **Aydin Boyaci**
Zum Stabilitäts- und Bifurkationsverhalten hochtouriger Rotoren in Gleitlagern. 2012
ISBN 978-3-86644-780-6

Band 16 **Rugerri Toni Liong**
Application of the cohesive zone model to the analysis of rotors with a transverse crack. 2012
ISBN 978-3-86644-791-2

Band 17 **Ulrich Bittner**
Strukturakustische Optimierung von Axialkolbeneinheiten. Modellbildung, Validierung und Topologieoptimierung. 2013
ISBN 978-3-86644-938-1

Band 18 **Alexander Karmazin**
Time-efficient Simulation of Surface-excited Guided Lamb Wave Propagation in Composites. 2013
ISBN 978-3-86644-935-0

Band 19 **Heike Vogt**
Zum Einfluss von Fahrzeug- und Straßenparametern auf die Ausbildung von Straßenunebenheiten. 2013
ISBN 978-3-7315-0023-0

Band 20 **Laurent Ineichen**
Konzeptvergleich zur Bekämpfung der Torsionsschwingungen im Antriebsstrang eines Kraftfahrzeugs. 2013
ISBN 978-3-7315-0030-8

Band 21 **Sietze van Buuren**
Modeling and simulation of porous journal bearings in multibody systems. 2013
ISBN 978-3-7315-0084-1

Band 22 **Dominik Kern**
Neuartige Drehgelenke für reibungsarme Mechanismen. 2013
ISBN 978-3-7315-0103-9

Band 23 **Nicole Gaus**
Zur Ermittlung eines stochastischen Reibwerts und dessen Einfluss auf reibungserregte Schwingungen. 2013
ISBN 978-3-7315-0118-3

Band 24 **Fabian Bauer**
Optimierung der Energieeffizienz zweibeiniger Roboter durch elastische Kopplungen. 2014
ISBN 978-3-7315-0256-2

Band 25 **Benedikt Wiegert**
Nichtlineare Schwingungen von Systemen
mit elastohydrodynamischen Linienkontakten. 2015
ISBN 978-3-7315-0350-7

Band 26 **Arsenty Tikhomolov**
Analytische, numerische und messtechnische
Untersuchung der Dynamik von Fahrzeugkupplungen
am Beispiel des Trennproblems. 2015
ISBN 978-3-7315-0362-0

Band 27 **Daniel Maier**
On the Use of Model Order Reduction Techniques
for the Elastohydrodynamic Contact Problem. 2015
ISBN 978-3-7315-0369-9

Band 28 **Xiaoyu Zhang**
Crosswind stability of vehicles under nonstationary
wind excitation. 2015
ISBN 978-3-7315-0376-7

Band 29 **Jens Deppler**
Ein Beitrag zur viskoelastischen Modellierung nichtholonomer
Bindungsgleichungen. 2017
ISBN 978-3-7315-0548-8

Band 30 **Georg Jehle**
Zur Modellbildung und Simulation reibungserregter
Schwingungen in Pkw-Schaltgetrieben. 2017
ISBN 978-3-7315-0668-3

Band 31 **Joachim Klima**
Lubricant transport towards tribocontact in capillary
surface structures. 2018
ISBN 978-3-7315-0814-4

Band 32 **Gábor Kenderi**
Nonparametric identification of nonlinear dynamic systems. 2018
ISBN 978-3-7315-0834-2